Statistical
Case Studies

ASA-SIAM Series on
Statistics and Applied Probability

The ASA-SIAM Series on Statistics and Applied Probability is published jointly by the American Statistical Association and the Society for Industrial and Applied Mathematics. The series consists of a broad spectrum of books on topics in statistics and applied probability. The purpose of the series is to provide inexpensive, quality publications of interest to the intersecting membership of the two societies.

Editorial Board

Statistical Case Studies

A Collaboration Between Academe and Industry

Student Edition

Roxy Peck
California Polytechnic State University
San Luis Obispo, California

Larry D. Haugh
University of Vermont
Burlington, Vermont

Arnold Goodman
University of California
Irvine, California

Society for Industrial and Applied Mathematics
Philadelphia, Pennsylvania

ASA
American Statistical Association
Alexandria, Virginia

Library of Congress Cataloging-in-Publication Data

Statistical case studies : a collaboration between academe and industry / [compiled by]
  Roxy Peck, Larry D. Haugh, Arnold Goodman.
  -- Student ed.
      p. cm. -- (ASA-SIAM series on statistics and applied
  probability)
   Includes bibliographical references and index.
   ISBN 0-89871-421-4 (pbk.)
  1. Statistics--Case studies.   I. Peck, Roxy. II. Haugh, Larry D. III. Goodman, Arnold.
  IV. Series.
  QA276.12.S734 1998
  519.5--dc21   98-17931

 Partial support for this work was provided by the Undergraduate Faculty Enhancement program at the National Science Foundation under DUE 9455055. The work is that of the authors and does not necessarily express the views of the NSF.

In memory of Joyce Curry-Daly, statistics educator extraordinaire.
Roxy Peck

For my wonderful family: Janie, Wendi, Josh, and Jeremy.
Larry Haugh

Honoring Herman Chernoff's distinguished contributions
and unique personal style on his recent retirement:
he shaped my thinking that led to Interfaces
and his Faces were 20 years ahead
of the computer visualization.
Arnie Goodman

Contents

PREFACE

If you only have pretend data, you can only pretend to analyze it.
George Cobb

As professors of statistics we tell our students that an understanding of research questions is necessary in order to collect meaningful data and analyze it intelligently. "Don't collect data first and then try to figure out what (if anything) you can do with it," we admonish students and also researchers who come to us for help with data analysis. Yet when we teach statistics courses, we often do just that! Now convinced of the necessity to include examples that use REAL data, we search for real data sets and then try to come up with some question that we think might in some way capture our students' interest. Without an in-depth understanding of how the data was collected or why it is important, we look at data and try to figure out what (if anything) we can do with it to turn it in to a classroom example. I confess that I am guiltier of this than most—I have created whole textbooks full of examples and exercises in just this way.

While examples and case studies of this type are certainly better than those based on artificial data and contrived situations, I now realize that a case study based on real data from industry that has research questions and directed analyses constructed by an academic often looks very different from a case study based on actual practice in industry. Traditional statistics education has been criticized for not adequately preparing statisticians and engineers for careers in industry and for being unresponsive to the needs of industry. This may be partially due to the fact that most university faculty do not have the type of industry experience that would enable them to easily incorporate examples based on actual practice in industry into their courses. We hope that this collection of case studies can help make this easier.

The collection of cases is eclectic—they come from a variety of application areas and the vast majority require the use of multiple data analysis methods. Some are challenging and some are messy—we have made no effort to "simplify" the problems presented for classroom use. What unifies this collection is that all are based on actual practice in industry or government.

Each case study in this collection is the product of a collaboration between statisticians in industry and colleagues in academe. These collaborations were made possible by support from the National Science Foundation's Division of Undergraduate Education through an Undergraduate Faculty Enhancement grant. Forty-four statisticians, 22 from academic institutions and 22 from business, government, and industry, participated in the Collaboration Project and in the development of these case studies. All participants met at a workshop held at Cal Poly, San Luis Obispo, during the summer of 1995. The workshop program focused on academe–industry partnerships and included presentations by Ron Iman, Bill Parr, Bob Mason, and Dick Gunst. Academe–industry pairs were also formed during this workshop. During the following academic year, each pair participated in three-day exchange site visits. These site visits enabled each participant to experience the work environment of his or her partner and became the basis for the development of the case studies that make up this collection. Twenty-one of the 22 pairs produced a case study (one pair produced three), and 20 of the 24 cases submitted are included in this volume.

The result of these individual collaborations is a collection of case studies that illustrate real application of statistical methodology to solve problems in industry. They address problems important to industry, as opposed to an academic view of what might be important. As a collection, they show the scope and variety of problems that can be addressed using statistical methods, and illustrate a number of points that are important for students of statistics:

- Real problems almost always require the sequential use of several statistical methods.
- There is almost always more than one reasonable answer.
- Problems encountered in actual practice are generally much messier than traditional *textbook* problems. Not all problems are going to look just like one from the text.

The collection can be used in several ways: (1) Statistical consulting courses are becoming more common in statistics curricula around the country. This collection could easily serve as a text for such a course. (2) An instructor could use selected cases to enrich just about any statistics course. (3) Practitioners of statistics in industry may find the collection a useful reference, in that it illustrates the application of a wide variety of statistical methods in applied settings.

The cases are ordered roughly by level of sophistication. Tables indexing the cases by statistical methods required (Table 1), by application area (Table 2), and by level of difficulty (Table 3) follow this preface.

Two introductory articles that will also be of interest to instructors and practitioners precede the collection of cases. The first, "The Benefits of Cases," by Bill Parr, examines how statistical case studies benefit a variety of stakeholders, including students, faculty, and statistics professionals. The second article, "Partnering for the Future of the Statistics Profession," by Ron Iman, looks at the long-range benefits of academe/industry collaboration.

ACKNOWLEDGMENTS

Many people contributed to the success of the Collaboration Project and to the completion of this collection of case studies. We were truly fortunate to have an outstanding group of 44 workshop participants who brought humor, dedication, and patience to the project. Bob Mason, Dick Gunst, Bill Parr, and Ron Iman, the featured workshop speakers, created a workshop program that both informed and motivated, and Rick Rossi helped with local arrangements. Barb Barnet, Vera Boulaevskaia, Gina Gustke, and Aaron Kaufman, students in statistics programs at Iowa State and the University of Minnesota, reviewed the initial drafts of the case studies and offered comments and suggestions from a student perspective. Technical support and assistance with manuscript preparation was provided by John Groves and Alicia Beckett. This collection would not have been possible without their help. And finally, we thank the National Science Foundation for its support of this project.

Roxy Peck
San Luis Obispo, CA

Table 1. *Index of case by statistical method.*

Chapter #	Title	Sampling	Experimental Design	Descriptive Statistics	Probability Distributions	Linear Regression, Correlation	Nonlinear Regression	Transform -ing Data	Logistic Regression
1	Are the Fish Safe ...	X							
2	Chemical Assay Validation					X			
3	Automating a Manual ...								
4	Dissolution Method ...	X							
5	Comparison of Hospital ...		X	X					
6	Comparing Nonsteriodal ...								
7	Validating an Assay ...					X			
8	Control Charts for Quality ...				X			X	
9	Evaluation of Sound ...								
10	Improving Integrated Circuit ...	X	X						
11	Evaluating the Effects of ...								
12	Designing an Experiment ...		X						
13	Investigating Flight Response ...		X						X
14	Estimating Biomass of ...	X		X			X		
15	A Simplified Simulation of ...			X	X				
16	Cerebral Blood Flow ...					X			
17	Modeling Circuit Board Yields								
18	Experimental Design for ...		X	X					
19	An Evaluation of the Process ...			X					
20	Data Fusion ...								

Table 1. *Continued.*

Chapter #	Title	Multiple Regression	Response Surface Methods	Estimation	Bias	Two Sample Comparisons	Power	Analysis of Variance	Multiple Comparisons
1	Are the Fish Safe …			X			X	X	
2	Chemical Assay Validation			X	X		X		
3	Automating a Manual …					X			
4	Dissolution Method …					X	X		
5	Comparison of Hospital …	X				X	X		
6	Comparing Nonsteriodal …								X
7	Validating an Assay …							X	
8	Control Charts for Quality …								
9	Evaluation of Sound …								
10	Improving Integrated Circuit …	X		X					
11	Evaluating the Effects of …		X						
12	Designing an Experiment …	X	X				X		
13	Investigating Flight Response …			X					
14	Estimating Biomass of …								
15	A Simplified Simulation of …								
16	Cerebral Blood Flow …	X						X	
17	Modeling Circuit Board Yields …							X	
18	Experimental Design for …							X	
19	An Evaluation of the Process …			X					
20	Data Fusion …								

Table 1. *Continued.*

Chapter #	Title	Paired Comparisons	Analysis of Covariance	Nonparametric Methods	Goodness of Fit	Statistical Process Control	Simulation	Time Series	Bayesian Analysis
1	Are the Fish Safe ...			X					
2	Chemical Assay Validation								
3	Automating a Manual ...							X	
4	Dissolution Method ...								
5	Comparison of Hospital ...								
6	Comparing Nonsteriodal ...			X					
7	Validating an Assay ...								
8	Control Charts for Quality ...				X	X			
9	Evaluation of Sound ...	X		X					
10	Improving Integrated Circuit ...								
11	Evaluating the Effects of ...						X		
12	Designing an Experiment ...					X			
13	Investigating Flight Response ...								
14	Estimating Biomass of ...								
15	A Simplified Simulation of ...						X	X	
16	Cerebral Blood Flow ...		X						
17	Modeling Circuit Board Yields								
18	Experimental Design for ...								
19	An Evaluation of the Process ...					X	X		
20	Data Fusion ...								X

Table 2. *Index of case by application area.*

Chapter #	Title	Biology/ Environment	Medical & Health Care	Pharmaceutical	Marketing & Survey Research	Manufacturing
1	Are the Fish Safe ...	X				
2	Chemical Assay Validation			X		
3	Automating a Manual ...				X	
4	Dissolution Method ...			X		
5	Comparison of Hospital ...		X			
6	Comparing Nonsteriodal ...		X			
7	Validating an Assay ...		X			
8	Control Charts for Quality ...					X
9	Evaluation of Sound ...				X	
10	Improving Integrated Circuit ...					X
11	Evaluating the Effects of ...				X	
12	Designing an Experiment ...					X
13	Investigating Flight Response ...	X				
14	Estimating Biomass of ...	X				
15	A Simplified Simulation of ...					X
16	Cerebral Blood Flow ...		X			
17	Modeling Circuit Board Yields					X
18	Experimental Design for ...					X
19	An Evaluation of the Process ...					X
20	Data Fusion ...					X

Table 3. *Index of case by course level.*

Chapter #	Title	Introductory (First Course)	Intermediate (Second Course)	Advanced Undergraduate or M.S.	Graduate M.S. or Ph.D.
1	Are the Fish Safe ...		X		
2	Chemical Assay Validation	X			
3	Automating a Manual ...		X		
4	Dissolution Method ...	X			
5	Comparison of Hospital ...		X		
6	Comparing Nonsteriodal ...		X		
7	Validating an Assay ...		X		
8	Control Charts for Quality ...		X		
9	Evaluation of Sound ...		X		
10	Improving Integrated Circuit ...		X		
11	Evaluating the Effects of ...		X		
12	Designing an Experiment ...		X		
13	Investigating Flight Response ...		X		
14	Estimating Biomass of		X		
15	A Simplified Simulation of ...		X		
16	Cerebral Blood Flow ...		X		
17	Modeling Circuit Board Yields		X		
18	Experimental Design for ...			X	
19	An Evaluation of the Process ...			X	
20	Data Fusion ...				X

THE BENEFITS OF CASES

Dr. William C. Parr

Why cases?

The question is simple. Amplified, it becomes: Why should we hope to see more cases of statistical applications written up? The purpose of this note is to provide an answer (not the only one) to that question.

My method is simple: I look at the question in terms of the interest of a variety of stakeholders:

- the student who takes a class using the case or reads the case,
- the organization which eventually hires the student,
- the faculty who teach using the case,
- the statistical profession,
- the company/organization in which the work documented in the case took place,
- the writer of the case.

We find that all of these stakeholders have a strong, positive interest in cases being written. Hence, win–win relationships built around purposeful case writing seems both mutually desirable and possible.

THE INTEREST OF THE STUDENT

Students clamor for real experience. Study after study validates that learning is best accomplished by doing. Internship and co-op programs prosper at ever-increasing levels, with employers in some cases not wanting to consider applications from students with no internship or co-op experience. (At the University of Tennessee Department of Statistics, we consistently find that when a student (rarely) foregoes an internship, they are typically among the last to secure a good job.)

Learning by cases provides students with a means to have the experience vicariously. They can study a case, do their own analysis, make their own recommendations, discuss these in a case-discussion class or turn in written analysis and recommendations, and then hear what was actually done. Then, they have the opportunity to compare what was actually done (which may be by no means the best approach) with what they (the students) did, with the collaborative support of their classmates and professor.

Students can work individually (often a better method for validating and growing individual competence) or in teams (excellent for growing teamwork skills, and for learning from each other).

Much like a flight simulator experience, the consequences of "failure" are fairly benign—if a student makes errors of analysis or judgment in recommendations, the only consequence is poor performance in the case. They do not lose major credibility inside their organization, lose millions of dollars for their company as a consequence of a poor decision, or otherwise have to live with the full consequences of their analysis. Instead, they learn from what others have to contribute, strengthen themselves and their understanding, and come again to the next class, prepared to contribute. Crashes are not fatal—they are valued learning experiences.

A further benefit of learning by case-method teaching is the ability of the student to participate in a give-and-take which can actually be comparable to what they will experience in the real work environment after graduation.

A further benefit of learning by case-method teaching is the ability of the student to participate in a give-and-take which can actually be comparable to what they will experience in the real work environment after graduation.

THE INTEREST OF THE HIRING ORGANIZATION

Hiring organizations want experience. They prefer students with internship and co-op experience. They want students who have had to continually look at situations, size them up, determine appropriate data collection strategy, analyze the data, and make and defend their recommendations.

Cases have several benefits over internships or co-op experiences. (Incidentally, we strongly support internships and co-ops, but recognize their limitations from the point of view of offering a variety of substantive applied experiences to students.)

One benefit is that a student can handle one major case a week, or even more, via case-based learning. However, in an internship, they would not be likely to see such a broad number and range of applications as can be examined through a judicious choice of cases. The result is that the hiring organization can hire a student who has been through 15 or more virtual (case-based) experiences per course, for a total of perhaps 100 or more in their curriculum. This experience is far different from that of working homework problems, involving a higher level of integration of statistical thinking into real technical, process, and business issues and problems.

There is of course incompleteness to the case analysis and case discussion experience. No case analysis and case discussion can duplicate fully the political environment in which improvement work takes place. This knowledge must be provided for the student by some other means (and internships and co-ops are excellent at providing this knowledge).

THE INTEREST OF THE TEACHING FACULTY

Teaching faculty want to teach material which is seen to be relevant. In many cases, teachers of statistics do not have active personal work in applied statistics. Some are not even statisticians, but instead are mathematicians by training who teach statistics as a part of their teaching load. Others were theoretically trained and have little or no consulting experience, and would be hard put to draw applied examples from their own experience.

Cases provide a way for faculty to draw interesting, meaty exercises in the use of statistical thinking from an experience range broader than the current instructor—one as broad as the set of all writers of statistical cases.

Cases provide a way for faculty to make clear the kinds of applications where statistical thinking can be useful. Cases can provide a way for faculty to imbed the statistical tools inside a broader problem solving and system improvement process.

THE INTEREST OF THE STATISTICAL PROFESSION

The statistical profession is in need of further documenting the value it provides. In the excellent book "Statistical Case Studies for Industrial Process Improvement," edited by Veronica Czitrom and Patrick D. Spagon, the editors begin their preface by saying that "The primary objective of this book is to demonstrate how American industry would benefit from using statistical methods." They continue, "Another major objective is to provide examples of successful industrial applications of statistics for use in industrial workshops and in academic courses."

THE INTEREST OF THE COMPANY/ORGANIZATION IN WHICH THE WORK WAS DONE

Many companies/organizations are interested in making their good work known. They see the value in publicizing this successful work. It can be an aid to them when they must hire statisticians in the future. Statisticians in academe can easily envision this effect—imagine what organizations you would recommend to a bright young student considering future employment. Then, consider whether these are organizations which have tended to make their applications of statistics relatively "secret" or instead those organizations which have publicized their experiences using statistical thinking.

THE INTEREST OF THE WRITER OF THE CASE

Traditionally, academics have found it difficult to place a positive value on applied work. (The author remembers one faculty discussion at an institution which will not be named in which someone who was fond of "counting publications" suggested that a tenure and promotion committee should not only count the publications but should subtract something based on the amount of consulting and other applied engagements—since they were, in his view, evidence of lack of a full commitment to the "correct" academic model. The eyes of the speaker did not twinkle.) The writing of cases gives those actively engaged in the practice of statistics a way to create written scholarly output which can be a natural outgrowth of their applied work, and yet meet the needs for scholarly output which are articulated by their academic institutions.

Much work is needed on this front. In some institutions, case writing is not viewed as a significant component of scholarly work.

A FEW WORDS ON GETTING STARTED USING CASES

The author would be remiss if he were not to give some indication of how to get insights into the use of cases. The literature on the use of cases continues to grow. One particularly useful source of cases, in addition to the current volume, is the Czitrom and Spagon volume previously cited (a number of very useful cases contained therein). A good source of information on teaching by the case method (not particular to statistics teaching) is Christensen, C. Roland, Garvin, David A., and Sweet, Ann (1991). Parr, W. C. and, Smith, Marlene (1997) have written on their experiences with teaching statistics by the case method.

Cases are used in many ways in the classroom—from constituting the basis of examples to document and illustrate applicability of methods to being the basis for case discussion in class, in which different students have individually analyzed the data and constructed their recommendations, and describe, advocate, and defend them in class. Each way has its own special benefits. We invite you to review the sources listed above, use the cases in this book, and learn from that practice.

CONCLUSION

We applaud the authors of these cases for their contributions to the growing statistical case literature. They toil in a vineyard which still looks for more workers—the deficit of statistical cases is appalling when one compares it to the widespread and growing use of statistical thinking to address real and significant problems. We also thank Roxy Peck for giving us the opportunity to be one of the faculty instructors for the original National

Science Foundation workshop which led to the writing of these cases, and for the opportunity to write this foreword.

REFERENCES

Christensen, C. Roland, Garvin, David A., and Sweet, Ann (editors) (1991), *Education for Judgment: The Artistry of Discussion Leadership*. Harvard Business School Press.

Czitrom, Veronica and Spagon, Patrick D. (editors) (1997), *Statistical Case Studies for Industrial Process Improvement*. SIAM, Philadelphia.

Parr, W. C. and Smith, Marlene (1997), Use of Cases in Teaching Statistics. Proceedings of Making Statistics More Effective in Schools of Business. Submitted to *American Statistician*.

BIOGRAPHY

William C. Parr is a Professor and former Head of the Department of Statistics at the University of Tennessee. Bill has extensive practical experience working as a consultant to industry and was a full-time Senior Scientist in charge of Statistical Development at the Harris Corporation, Semiconductor Sector. At the Harris Corporation, he managed an organization of statisticians and support staff providing consulting and training to an organization of roughly 12,000 employees. Bill was closely involved with efforts at Harris Corporation to implement the philosophy of statistical management, which led to presenting the 1984 Institute for Productivity through Quality Achievement Award to the Semiconductor Sector. In addition, he consulted with senior management on the management and cultural implications of statistical tools and the philosophy of continuous improvement. He has published over 30 papers and does research in the areas of management, statistical management, and systems analysis. A special area of research interest over the last few years has been the use of cases in teaching statistics.

Partnering for the Future of the Statistics Profession

Ronald L. Iman

Introduction

Many of us are all too aware that it is not business as usual anymore; competition is forcing changes. We see changes in the way organizations are managed—witness the breakup of AT&T into three separate entities. The statistics profession is not immune to such changes—after all, software makes anyone a statistician! If the statistics profession is to remain relevant, we must make major changes in the operation of our professional organizations, in the way we train our students, in the way we interact with our customers, and in the way we perceive our profession.

The problems we face today are too complex for any one entity to solve in isolation, and in this era of increasingly tighter budgets, academe and industry are seeking ways to leverage scarce resources. The statistics profession should flourish in such an environment! However, statisticians seem all too willing to take a back seat while others make decisions affecting their profession. We seem content to teach decision making rather than educating the decision makers who control our livelihood.

What can we do in the current situation? For one thing, we can learn from and emulate the success of others. For example, industry is placing greater emphasis on partnering with their suppliers and customers. These partnership efforts have decreased production costs, shortened production time, and increased product performance. The statistics profession could benefit by partnering with other organizations having a professional interest in statistics. Currently, there is little or no interaction among such organizations. When was the last time two or more professional organizations held a joint meeting to see how they might work together to benefit the statistics profession?

The statistics profession would also profit through partnering activities between academe and industry/government. However, the concept of partnering is an enigma to most universities as their administrative structures, with rigidly defined departments and colleges, encourage an inward focus. Few universities recognize the existence of a customer–supplier relationship either within the university or in their relationships with external organizations. This inward focus has had a pernicious effect not only on universities, but also on industry. University departments evaluate research in comparison with peer institutions and frequently have a low regard of joint research, rather than evaluating it in terms of its usefulness by industry. Traditional training methods have not emphasized the need for cross-disciplinary training and teamwork. Consequently, graduates are not equipped to work as team members and industry frequently has to provide special training to its new employees to ensure they speak each other's language. Statistics programs are usually evaluated by peer institutions without input from industry. As a result, industry is left on the outside looking in and universities wonder why funding from industry and government is increasingly difficult to obtain.

SEMATECH: A ROLE MODEL FOR PARTNERING

When I was President of the American Statistical Association in 1994, I asked Dr. Bill Spencer, the President and CEO of SEMATECH in Austin, to give the President's Invited Address at the ASA Annual Meeting in Toronto. I asked Dr. Spencer to talk about the success that SEMATECH achieved through partnering to help the U.S. regain its world leadership position in both semiconductor sales and semiconductor equipment sales. This success did not happen by accident nor did it come about easily. However, I believe the SEMATECH model is one that we can successfully implement in the statistics community. This model provides important lessons for the statistics community about the value of working together and thereby increasing our circle of influence.

The formation of SEMATECH was a difficult process since it involved getting competitors in the semiconductor industry such as Advanced Micro Devices, AT&T, Harris Corp., Intel, IBM, Motorola, and Texas Instruments, among others, and the federal government to work together. These manufacturers were in direct competition with one another for what amounted to an ever-decreasing share of the worldwide semiconductor market.

Many doubted that SEMATECH would be able to succeed in a nation where fierce competition flourished and industry cooperation was minimal. Moreover, the government had been a roadblock in allowing these companies to work together. However, these companies and the government quickly learned that they had more to gain by working together and leveraging their resources than by trying to go it on their own.

SEMATECH had another big obstacle to overcome involving customer–supplier relationships. SEMATECH has a large group of suppliers that is collectively known as SEMI/SEMATECH. The members of this latter group range from very small to very large and they supply a wide spectrum of materials used in semiconductor manufacturing. In turn, this group of suppliers has their own group of suppliers. SEMI/SEMATECH took the lead in forming partnerships between customers and suppliers. Again, many doubted that these partnerships would work, and each side approached them cautiously. However, these partnerships have made both parties stronger, improved the product, reduced costs, and are now regarded as the standard way to do business. They have learned that the whole is truly greater than the sum of the parts.

In 1979, the respective U.S. and Japanese shares of the worldwide semiconductor market were approximately 58% and 26%. By the time SEMATECH was formed in 1987, the U.S. share had declined to 38% while Japan's share increased to 48%. The following year, the U.S. share declined to 32% and Japan's share increased to 52%. As a result of SEMATECH's efforts, the U.S. pulled even with Japan in 1992 at a 42% share and regained the lead in 1993.

SEMATECH's success serves as a truly remarkable example of what can happen when competitors work together and customer–supplier relationships are openly recognized and cultivated. I stated in my ASA Presidential Address that many in our profession have pointed out that we do a poor job of recognizing the existence of customers. Thus, we must begin by recognizing that each of us has customers and that ours is a customer-driven profession.

SOME EXISTING PARTNERSHIPS IN THE STATISTICS PROFESSION

I chaired the opening session in Toronto during which four speakers gave presentations on the *Benefits of Increasing Interaction Among Statisticians in Academia and Industry*.

The speakers in this session were Bob Hogg from the University of Iowa, John Schultz from The Upjohn Company, Bob Mason from the Southwest Research Institute, and Ray Myers from VPI and State University. These speakers did an outstanding job of covering the need and identifying the benefits of increased interactions from a variety of viewpoints—academe, industry, and government. They made it clear that there are many advantages for all parties resulting from increased interactions. Moreover, they identified many of the changes that are needed to bring about increased interactions.

John Schultz provided an outstanding example of a successful partnership that has been going on for years between The Upjohn Company and Western Michigan University. There are other examples of successful partnerships between academe and industry and between academe and government that we can use as role models. For example, Oakland University has a very successful program with Ford Motor Co. that has created a true win–win situation for all parties. The University of Manitoba is located in a small community that is isolated geographically. However, their faculty has conducted workshops for local small businesses and this has led to a very successful program for their department. It is encouraging to see such partnerships, but we need to address the issue of partnerships on a wide-scale basis so that the opportunity for participation is greatly expanded.

A REVIEW OF SOME MODELS FOR PARTNERING

If we were to make a list of concerns to the statistics profession, we would find that it is not much different from those developed by many other professions. For example, the following list of concerns apply broadly to many professions:

- Lack of recognition
- Misrepresentation in the news media
- Journals are not readable
- Academe ignores industry's needs
- Joint research is not highly valued
- Misuse of statistics by nonprofessionals
- Skills or graduates are too narrowly focused
- Lack of jobs
- Shortage of funding
- Difficult to find students

In light of these concerns, I have defined three models that can be used to describe existing relationships between academe, industry, and government. I call the first model the Inward Focus Model. It is shown graphically in Figure 1. Academic products in this model are identified as students and research. However, customers for these products are viewed as academe while industry and government are viewed as a source of funding to support the system.

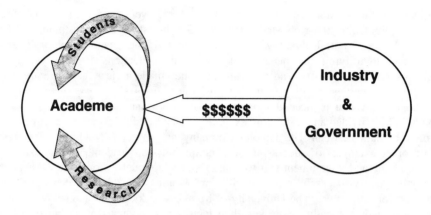

Fig. 1. *The Inward Focus Model.*

The following summary describes the characteristics of the Inward Focus Model as viewed by academe and industry.

Model Characteristics: Academe

- Research oriented to peers
- Our students will all teach at Berkeley
- Give us money and we will see what we can do
- Program review performed by peer institutions
- Inward focus is rewarded
- What's a customer?

Model Characteristics: Industry

- Academic research is not relevant to our needs
- Graduates are not trained in non-technical skills
- Academe only wants our financial support
- Seldom asked to participate in program reviews
- Academe is an ivory tower
- We are your customers!

The Inward Focus Model clearly does not bode well for establishing or maintaining healthy long-term relationships. We might ask if its focus on peer institutions is either realistic or justified. A partial answer to this question can be obtained by comparing the number of Ph.D. graduates in statistics over a 21-year period with the number of available faculty positions. As shown in Figure 2, the number of Ph.D. graduates in statistics has varied from approximately 173 to 327, with an average of 233. There were approximately 735 statistics faculty in the U.S. in 1990 with a growth rate for the preceding 21 years of approximately 1.75 new positions per year. Using a tenure figure of 30 years for a faculty position gives 735/30 = 24.5 positions available each year through retirements. Thus, there are only enough faculty positions available each year for about 10% of the new Ph.D. graduates. Or, viewed from the other perspective, 90% of the Ph.D.s will work outside of academe, and this does not account for the B.S. and M.S. graduates that will most likely be working in industry or government! These figures make it clear that the Inward Focus Model does not meet the needs of society and does not enhance the statistics profession.

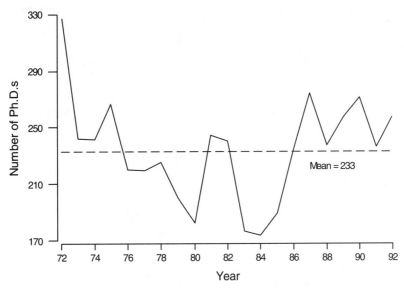

Fig. *2. Number of Ph.D. graduates in statistics from* 1972 *to* 1992.

The *Modified* Inward Focus Model is represented in Figure 3. Unlike the Inward Focus Model, this modified model recognizes the existence of an external customer, but it still has a heavy focus on peer institutions. The following summary describes the characteristics of the Modified Inward Focus Model as viewed by academe and industry.

Model Characteristics: Academe

- Academic sabbaticals in industry
- Student summer internships
- Some joint research activities
- Some industry adjoint professors
- Some funding from industry
- Some equipment from industry

Model Characteristics: Industry

- Industry adjoint professors
- Student summer internships are beneficial
- Joint research is helpful
- Funding of research provides some benefits
- Good re-use of equipment
- Limited input into academic programs

STATISTICS PARTNERSHIPS AMONG ACADEME, INDUSTRY, AND GOVERNMENT

As President-Elect-Elect of the ASA in 1992, I attended separate meetings of the Academic Program Representatives and the Corporate Member Representatives held during the Joint Statistical Meetings in Boston. I suggested that these groups try something "bold" and have a joint meeting. The two groups heeded my advice and held a joint meeting in San Francisco in 1993. The idea for summer internships came out of that meeting. This is now a yearly program that has been extremely helpful to all parties. At the annual ASA meeting held in Toronto in 1994 I again challenged the two groups to form strategic partnerships known as Statistics Partnerships among Academe, Industry, and Government (SPAIG). I asked each of the groups to independently develop vision statements, recommendations for achieving those visions, and related issues and concerns. I then suggested that they exchange their statements to see each other's views. These summaries were presented at a joint meeting of the two groups at the annual ASA meeting in Orlando in 1995. I now present a summary of the statements prepared by the two groups.

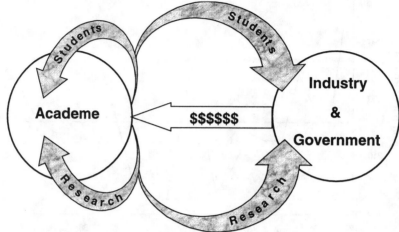

Fig. 3. *The Modified Inward Focus Model.*

The Modified Inward Focus Model is an improvement over the Inward Focus Model, but it still falls short of meeting the needs of the statistics profession. The last, and best, model is the Partnering Model. This model is represented in Figure 4.

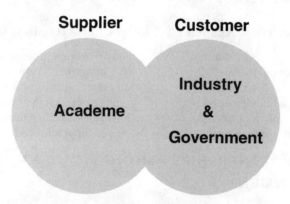

Fig. 4. *The Partnering Model.*

The characteristics of the Partnering Model can be summarized as follows:

- True customer–supplier relationship exists
- Training is driven by industry needs
- Joint research is highly valued
- Interactions with industry provide a source of real problems for Ph.D. students
- Long-term funding received from industry
- Students are highly sought as employees

- Industry participates in program reviews
- Two-way employee exchanges are the norm
- Research supports industry needs
- Equipment received from industry
- Student internships are the norm
- Enrollments are up
- The whole is greater than the sum of the parts
- Recruitment costs are lower and more effective

The Partnering Model clearly creates a win–win situation for all parties.

The View from Academia

Professor Bill Parr from the University of Tennessee took the lead in developing statements for the academic community. Professor Parr used e-mail to send drafts of the academic statements to department heads and statistics programs. This process was repeated five times with an average of over 40 responses each time. The very forward looking statements that resulted from that process are as follows.

If we are successful in cooperation between industry (includes both industry and government) and academia, in ten years we will be able to truthfully say the following:

In the area of exchange:
*   Statisticians from academia and industry routinely exchanges roles, with faculty taking one-semester to one-year "sabbaticals" from academia to contribute to industry, and industrial statisticians taking one-semester to one-year "sabbaticals" to work within universities.

In the area of education:
*   We (in academia) know what industrial organizations want, and we supply statisticians with the knowledge, skills, and attitudes to be effective both when they are initially employed and throughout their careers (through lifelong learning).
*   We have real examples and applications to present in the classroom, based on what we have learned about industrial practice.
*   We send out students routinely for internships and co-op arrangements, which benefit industry in hiring decisions, students in better value added to their education, and academics in more input for relevance. These internships and co-op arrangements have forged strong linkages between us and industry.
*   We seek out industrial statisticians to participate with us by giving seminars to our departments and helping our students and faculty understand the current realities of industrial practice.

In the area of research:
*   We are able to guide our research based on the needs of actual users of statistics.
*   We pursue industry as a developmental funding source.
*   We lead industrial problem solving into the future with new developments in theory and the means to apply that theory in industrial settings, appropriately targeted to meet real needs.
*   We transfer new knowledge in multiple ways, via courses, seminars, workshops, consulting arrangements, outgoing students, and the written (and electronic) word.
*   We have worked through current prohibitions against proprietary research to facilitate ongoing working relationships between universities and industry (not only between university statisticians and industry).

In the area of academic values:
*   We believe that sending a Ph.D. graduate to a good industrial position requiring Ph.D.-level competence is as valuable to our program and the profession as sending her/him to an academic position.
*   We give as much credit for a joint publication addressing a significant real problem in a creative way as we do for a theoretical paper addressing a significant problem in a creative way.

- We use industry as laboratories for the development of new theory and applicable techniques.

What do we hope to see on behalf of industry?
- Industry informs academia of their needs for broadly educated statisticians rather than menu-driven technicians.
- Industry acts to energize the current batch of academicians teaching statistics to transforming both their content and their pedagogy based on best current practices.
- Industry supports professional societies and academic institutions in the transfer of new knowledge through seminars, workshops, and consulting arrangements.
- Industry uses the academy in the solution of industrial problems through research grants and contracts and collaborating with academic institutions in the establishment of research centers.

The View from Industry

The corporate view was coordinated by Bruce Rodda from Schering-Plough Research Institute. It is also very forward looking and complements the views of academe.

Statisticians beginning careers in an industrial setting often have difficulty in adapting to the dramatic change from the academic to the industrial environment. At the time the new employee enters their company, they may have spent as much as 20 years in an academic environment and are often unaware of the expectations and opportunities in an industrial setting. Although they may have excellent theoretical and technical skills, many find that these alone are not sufficient for an opportunity-rich and professionally satisfying career.

Many statisticians arriving in industry view themselves as technicians and analysts whose primary role is to support the work of other disciplines. Although these technical functions are extremely important, acceptance of them as long-term career goals reduces the magnitude of the contributions statisticians can make and will also limit their career opportunities. This limitation minimizes the impact that statistics can have since statisticians at higher positions in organizations can, in general, have a much greater strategic influence than those in lower level (primarily technical) positions.

Academia and industry have very similar values. Industrial statisticians clearly know and value the academic interest in theory and research, whereas the academic may view the real world use of statistics in the industrial setting as a productive consequence of statistical research and education. Thus, the complete education of a statistician is a shared responsibility between academia and industry. Historically, the primary source of theoretical training for the industrial statistician has been in their formal academic training. In contrast, the practical utilization of this training through real-world problems has been acquired in the on-the-job setting.

The basic education that students receive prior to graduation should include both the excellent technical training that they now receive and an even greater exposure to a broad variety of practical problems. In addition, an understanding of the industrial environment and career opportunities will be very beneficial in providing students the broadest of options for their career choices. This broader background would allow graduates the opportunity to choose any number of areas of statistical application and possess the fundamental scientific and leadership skills to function as principals in their organization.

A survey of several corporate member representatives of the ASA resulted in a number of suggestions to achieve this objective. These can be categorized into two areas:

a. Broader understanding of the needs, opportunities, and environments in which statisticians work in an industrial setting.

- There needs to be an appreciation for teams, organization, society, and their relationships and dynamics.
- Statisticians must understand and appreciate the complete research process.
- Statisticians must understand the expectations of industry and have the training to meet those expectations.
- Statisticians should be interested in the total problem, not just the statistical part.
- Statisticians should have a global view of a research effort and have a systems approach to research.
- Statisticians must expand their skills beyond statistics to include those of a successful consultant (e.g., communication, negotiation, project management).
- Statisticians must be integral to the project and participate to its conclusion.

b. More formal associations between academia and industry that demand a true collaborative approach to the education of both students and continuing education.

- Industrial seminar series at universities should occur regularly.
- Faculty/industry personnel exchange should occur regularly.
- Faculty sabbaticals in industry should occur regularly.
- Industrial advisory committees at universities should be common.
- Adjunct faculty appointments of industrial statisticians should be more frequent and routine.
- Formal training/requirements in writing and presentation skills should be established.
- Cooperative research endeavors should be developed.
- The availability and content of continuing education programs should be broadened.
- A syllabus of both statistical and essential nonstatistical skills required by industry statisticians should be developed and issued under the aegis of the ASA.
- Academia needs to have a formal mechanism for understanding and anticipating the needs of industry and government.
- Faculty must have a genuine interest in working on the projects arising in industry. This is key—and is necessary, but not sufficient.
- Good professional interaction between faculty members and one or more industrial researchers needs to be established.
- True success will require that both academia and industry gain more from the interaction than they can accomplish separately.
- Universities should routinely invite industry scientists and statisticians to give seminars on real problems and to play an important role in the education of new statisticians.

There is a very strong interest in the industrial community to contribute to the development and education of our students. The advantages that could be provided by a more intensive association between academia and industry would result in statisticians being ready to assume more effective positions in industry and provide a greater opportunity for statisticians to lead research efforts in the future.

CURRENT SPAIG ACTIVITIES

SPAIG held a two-day workshop in Raleigh, NC on May 30–31, 1997. The workshop was chaired by Dr. Robert Starbuck of Wyeth-Ayerst and was attended by 87 senior-level statisticians from academe, industry, and government. Dr. Starbuck opened the workshop by mentioning valuable opportunities that could be achieved by better and more widespread partnering relationships:

- Expanding and improving the use of statistics
- Increasing the value of statistics to society
- Improving the educational experience of students
- Improving the career decision process and outcome
- Increasing communications among all statisticians
- Enabling greater awareness of each other's needs, issues, and concerns
- Improving the self-image of statisticians
- Making statistics a more rewarding profession
- Ensuring that statistics is a growth field

The SPAIG workshop provided a valuable opportunity for interaction among the different groups. Details of the workshop can be found on the SPAIG web site at *http://funnelweb.utcc.utk.edu/~wparr/spaig.html*. Information on detailed action plans, milestones, responsibilities, measures of success, and progress as it occurs will be regularly posted to the site. One important event at the workshop occurred when participants were asked to identify the consequences of maintaining the status quo, i.e., doing nothing to improve the partnering relationships between academe, industry, and government. Their responses are summarized as follows:

Consequence	Number of Votes
1. Statistics as a discipline viewed as irrelevant; decline in influence	28
2. Nonstatisticians will do statistics	28
3. Decline or elimination of statistics departments and professional societies	19
4. Students not prepared to solve future problems	12
5. Slow growth in technical advances in statistics	12
6. Industry/government will do in-house statistics training	11
7. Reduced contribution of statistics to society	10
8. Failure to attract good students	9
9. Reduced dollars, resources, support	8
10. Good applied statisticians not available for hire	7
11. Fewer employment opportunities	5
12. Continued alienation among academe and government	5
13. Outside forces will determine the direction of the discipline	3
14. U.S. Industry will be less competitive	2
15. Miscellaneous	11

WHAT CAN YOU DO?

It will take much more than the efforts of just a few individuals to make partnerships between academe and industry a reality. Everyone needs to participate and get involved. There are a number of ways to do this, including the following:

- Participate in local chapter activities for ASA, ASQC, IIE (Institute of Industrial Engineers), and SPE (Society of Professional Engineers). Expand your chapter participation activities and do not restrict yourself to just your chapter.
- Faculty members should initiate discussions within their departments on the advantages of partnering with industry and how they might go about doing so.
- Industrial statisticians should initiate discussions within and between their organizations on how to better partner with academe.
- Universities should regularly invite industrial statisticians to give seminars on actual applications of statistics.
- Academe should go to industry and present seminars that are of interest to industrial statisticians (i.e., not theoretical).
- Publicize your activities by writing articles for newsletters of professional organizations.
- Organize sessions on industrial applications of statistics at regional and national meetings.
- Contact board members of professional organizations with your concerns and ideas.
- Support and participate in SPAIG activities.

BIOGRAPHY

Ronald L. Iman is President and founder of Southwest Technology Consultants in Albuquerque. Prior to forming STC, he was a Distinguished Member of the Technical Staff at Sandia National Laboratories in Albuquerque, where he was employed for more than 20 years. He holds a Ph.D. in statistics from Kansas State University, is the author of 6 college textbooks in statistics, and has more than 110 professional publications. He served as the 1994 President of the American Statistical Association, is a Fellow of the ASA, and received the ASA Founders Award. He is also the recipient of a Distinguished Service Award from Kansas State University and a Distinguished Alumni Award from Emporia State University.

CHAPTER 1

ARE THE FISH SAFE TO EAT? ASSESSING MERCURY LEVELS IN FISH IN MAINE LAKES

Jennifer A. Hoeting and Anthony R. Olsen

The information in this article has been funded in part by the United States Environmental Protection Agency. It has been subjected to Agency peer review and approved for publication. The conclusions and opinions are solely those of the authors and are not necessarily the views of the Agency.

Mercury is a toxic metal sometimes found in fish consumed by humans. The state of Maine conducted a field study of 115 lakes to characterize mercury levels in fish, measuring mercury and 10 variables on lake characteristics. From these data, we can investigate four questions of interest: 1. Are mercury levels high enough to be of concern in Maine lakes? 2. Do dams and other man-made flowage controls increase mercury levels? 3. Do different types of lakes have different mercury levels? 4. Which lake characteristics best predict mercury levels?

INTRODUCTION

In May, 1994, the state of Maine issued the following health advisory regarding mercury in Maine lakes, warning citizens of the potential health effects of consuming too much fish from Maine lakes [Bower et al., 1997]:

> "Pregnant women, nursing mothers, women who may become pregnant, and children less than 8 years old, should not eat fish from lakes and ponds in the state. Other people should limit consumption (eating) fish from these waters to 6–22 meals per year. People who eat large (old) fish should use the lower limit of 6 fish meals per year. People who limit themselves to eating smaller (younger) fish may use the upper limit of 22 fish meals per year."

This health advisory resulted in newspaper headlines throughout the state proclaiming, "Mercury: Maine Fish are Contaminated by this Deadly Poison" (*The Maine Sportsman*), "Maine's Most Lethal Sport" (accompanied by pictures of ice fishermen) (*Maine Times*), and "Natural-Borne Killer, Mercury Rising" (*Casco Bay Weekly*). Were these newspapers issuing bold headlines merely to increase circulation or were they repeating stories based on fact? The data described below can give us some answers to this question.

WHAT IS MERCURY?

Mercury is a "heavy metal" that occurs naturally in the environment in several forms (elemental, organic, and inorganic). Mercury occurs naturally in the earth's crust and oceans and is released into the earth's atmosphere. In addition, human activity results in releases of mercury through the burning of fossil fuels and incineration of household and industrial waste.

Mercury enters fish through two known mechanisms. When water passes over fish gills, fish absorb mercury directly from the water. In addition, fish intake mercury by eating other organisms. Mercury tends to bioaccumulate at the top levels of the food chain. Bioaccumulation occurs when microorganisms convert inorganic mercury into toxic organic compounds which become concentrated in fatty tissues as they move up the food chain [EPA, 1994].

Mercury is a toxin that acts upon the human nervous system. Consumption of mercury-laden fish can lead to a variety of neurological and physiological disorders in humans. Because mercury acts upon the nervous system, developing children and fetuses are especially sensitive to mercury's effects [Bahnick et al., 1994].

BACKGROUND INFORMATION

In 1993, the U.S. Environmental Protection Agency (EPA) and the state of Maine implemented the "Maine Fish Tissue Contamination Project." The goals of the project were to determine the distribution of selected contaminants in fish from Maine lakes, to determine risk to human and wildlife consumers of fish from Maine lakes, and to identify factors that affect the distribution of contaminants in fish tissue. To select the sample of lakes, the research team identified 1073 lakes in Maine that had previously been surveyed, were found to have significant fisheries, and were reasonably accessible. The identified lakes are a subset of the total number of lakes in Maine, 2314 [USEPA, 1995]. From the 1073 lakes, a simple random sample of 150 lakes was selected for study. Out of the original 150 lakes selected, samples were collected from only 125 of these lakes during the summers of 1993 and 1994. Nonsampled lakes were either not reasonably accessible or did not have desired fish species available.

A group of "target species" were determined based on the species' desirability as game fish and other factors. The data included here involve only the predator species from the original target species list and thus only 115 lakes out of the original list of 150 lakes are included (Fig. 1). To collect the fish specimens, field crews obtained up to 5 fish from the hierarchical order of preferred predator species group. Field protocols targeted fish that were of comparable age, legal length limit, "desirability" as game species, and likelihood of capture. Fish were collected by angling, gill nets, trap nets, dip nets, or beach seines. Care was taken to keep fish clean and free of contamination. Upon capture, fish were immediately killed if alive. Fish were rinsed in lake water and wrapped in aluminum foil, labeled with an identification number, and kept on ice in a cooler. Upon returning from the field, fish were immediately frozen for later analyses. In the laboratory, the fish fillet (muscle) of each fish was extracted. The fillets from each lake were ground up, combined and homogenized, and then the tissue was subsampled to analyze for mercury levels.

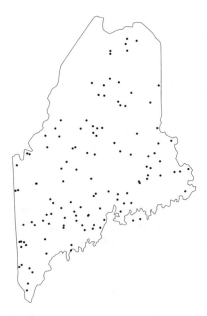

Fig. 1. *Sampled lake locations. Note: The points that appear to be in the Atlantic Ocean are located on islands.*

Another goal of the study was to examine external stressors and other factors potentially responsible for elevated levels of mercury in fish. The information would be used to gain insights on conditions and sources that could be used in managing any problems detected. The factors were divided into fish factors, lake factors, and geographic stressors (watersheds and airsheds). Only a subset of the factors are used here. Lake characteristics include lake size, depth, elevation, lake type, lake stratification, watershed drainage area, runoff factor, lake flushing rate, and impoundment class.

Some useful definitions

Lake type. This is a lake classification system [Collie, 1992]. A trophic state or level is a classification of taxa within a community that is based on feeding relationships. An *oligotrophic* lake has a balance between decaying vegetation and living organisms, where the lowest layer of water never loses its oxygen and the water contains few nutrients but sustains a fish population. A *eutrophic* lake has a high decay rate in the top layer of water and so contains little oxygen at the lowest levels; it has few fish but is rich in algae. A *mesotrophic* lake is between the oligotrophic and the richer eutrophic state and has a moderate amount of nutrients in its water.

Lake stratification. This refers to temperature stratification within a lake. In summer, the lake surface warms up and a decreasing temperature gradient may exist with the bottom remaining cold. Consider a lake stratified if a temperature decrease of 1 degree per meter or greater exists with depth.

Drainage area. This is the area of land which collects and drains the rainwater which falls on it, such as the area around a lake [Collie, 1992].

Runoff factor. RF = (total runoff during year)/(total precipitation during year). Runoff is the amount of rainwater or melted snow which flows into rivers and streams. In general, higher runoff factors may lead to more surface waters from the lake watershed reaching lakes. If contaminants are from local sources, this may influence concentrations in fish.

Flushing rate. FR = (total inflow volume during year)/(total volume of lake). This gives the number of times all water is theoretically exchanged during a year.

QUESTIONS OF INTEREST

1. The U.S. Food and Drug Administration has determined that samples with more than 1.0 ppm (parts per million) mercury are above the safety limit. Most states consider 0.5 ppm mercury levels (Maine uses 0.43 ppm) to be high enough to consider taking action (e.g., issuing a health advisory, considering methods of clean-up, etc.). As indicated by the data collected here, are mercury levels high enough to be of concern in Maine?
2. The industries that benefit from dams and dam construction are concerned that environmentalists will claim that high mercury levels in fish are related to the presence of a dam (or man-made flowage) in the lake's drainage. Do the data support this claim?
3. Previous studies [Nilsson and Håkanson, 1992; Larsson et al., 1992] suggest that mercury levels vary by lake type with oligotrophic lakes experiencing the highest mercury levels and eutrophic lakes experiencing the lowest mercury levels. Do the Maine data support this claim?
4. In future studies, it would be useful to predict mercury levels using lake characteristics as the latter are inexpensive data to collect. Which lake characteristics best predict mercury levels?

ADDITIONAL QUESTIONS FOR ADVANCED STUDENTS

- Do the missing data appear to be missing at random? If we omit the lakes with missing data from the analysis, how does this influence the resulting inferences?
- Should the number of fish per sample be taken into account in your regression analysis?

DATA

Name of data file: Case01.txt

Table 1. *Maine lake data.*

CODE	DESCRIPTION
NAME	Lake or pond name
HG	Mercury level in parts per million (ppm)
N	number of fish in the composite
ELV	elevation (feet)
SA	surface area (acres)
Z	maximum depth (feet)
LT	lake type as determined by the Department of Inland Fisheries and Wildlife 1 = Oligotrophic, 2 = Eutrophic, 3 = Mesotrophic
ST	lake stratification indicator (1=yes, 0=no)
DA	drainage area (square miles)
RF	runoff factor
FR	flushing rate (number flushes per year)

CODE	DESCRIPTION
DAM	Department of Inland Fisheries and Wildlife impoundment class 0 = no functional dam present; all natural flowage 1 = at some man-made flowage in the drainage area
LAT1	Latitude degrees
LAT2	Latitude minutes
LAT3	Latitude seconds
LONG1	Longitude degrees
LONG2	Longitude minutes
LONG3	Longitude seconds

Table 1 lists the variables available for the data analysis.

The first five lines of Case01.txt are shown in Table 2 as they appear in the file. In Case01.txt, missing data are indicated by "–9." Note: five lakes have duplicate names, but they are different lakes (see latitude and longitude columns).

Table 2. *First five lines of Case01.txt.*

N	NAME	HG	N	ELV	SA	Z	LT	ST
1	ALLEN.P	1.080	3	425	83	27	3	1
2	ALLIGATOR.P	0.025	2	1494	47	26	2	0
3	ABASAGUNTICOOK.L	0.570	5	402	568	54	2	1
4	BALCH&STUMP.PONDS	0.770	5	557	704	44	2	1
5	BASKAHEGAN.L	0.790	5	417	6944	22	2	0

Table 2. *First five lines of Case01.txt (continued).*

DA	RF	FR	DAM	LAT1	LAT2	LAT3	LONG1	LONG2	LONG3
2	0.60	2.8	1	44	57	44	68	5	7
1	0.69	0.8	1	45	37	50	69	12	30
15	0.56	1.1	0	44	25	13	70	19	22
14	0.58	2.7	0	43	37	0	70	59	4
123	0.57	2.0	1	45	30	32	67	50	2

INSTRUCTIONS FOR PRESENTATION OF RESULTS

Write a report addressing the questions above. Include summary statistics for each variable in relevant plots and tables. Interpret your results. You may want to seek out background material on mercury. The list of references below would be a good place to start your library search.

Assume the report will be read by Maine's governor and legislature. These individuals are concerned about the impact of mercury on the tourism and recreational fishing industries. Since Maine's governor and legislature may not have a background in statistics, make sure the main points in your report can be understood by a nonstatistician.

REFERENCES

Bahnick, D., C. Sauer, B. Butterworth, and D.W. Kuehl (1994), *A national study of mercury contamination of fish*, Chemosphere, 29(3):537–546.

Bower, Barry, Jeanne DiFranco, Linda Bacon, David Courtemanch, Vicki Schmidt, and John Hopeck (1997), *Fish Tissue Contamination in Maine Lakes*. DEPLW97-6. State of Maine Department of Environmental Protection, Augusta, Maine.

Collie, P. (1992), *Dictionary of Ecology and the Environment*, Collie: London.

Goad, Meredith (1994), "*State issues fish warning: Mercury levels prompt consumption advisory.*" Kennebec Journal, 19 May 1994, Section 1, p. 1.

Karr, Paul (1994), "*Natural-Borne Killer. Mercury Rising,*" Casco Bay Weekly, 3 November 1994, p. 1.

Larsson, P., L. Collvin, O. Lennart, and G. Meyer (1992), *Lake productivity and water chemistry as governors of the uptake of persistent pollutants in fish*. Environ. Sci. Technology, 26(2):346–253.

Nilsson, Å. and L. Håksanson (1992), *Relationships between mercury in lake water, water color and mercury in fish*, Hydrobiologia, 235/236:675–683.

Tracewski, Kevin (1995), "*Mercury: Maine fish are contaminated by this deadly poison. But how serious is the problem? Here are the best answers available.*" The Maine Sportsman, February 1995, p. 73.

U.S. Environmental Protection Agency (1994), *Summary Review of Health Effects Associated with Mercuric Chloride: Health Issue Assessment*, EPA/600/R-92/199, Washington, DC.

U.S. Environmental Protection Agency (1995), *National Water Quality Inventory: 1994 Report to Congress (and Appendixes)*, EPA/841/R-95/005 and EPA/841/R-95/006, Washington, DC.

Weegar, Andrew K. (1995), "*Maine's most lethal sport. The bigger the fish, the greater the risk of mercury poisoning. But nobody—not even the state—seems to care,*" Maine Times, 27 January 1995, p. 1.

BIOGRAPHIES

Jennifer A. Hoeting is Assistant Professor of Statistics at Colorado State University. She received her Ph.D. in 1994 and M.S. in 1991 from the University of Washington Department of Statistics. Her research interests include Bayesian statistics, linear models, statistical computing, and statistical methodology for environmental problems. In addition to her more traditional professorial roles as academic researcher and instructor at CSU, Hoeting serves as project supervisor for the Department of Statistic's Center for Applied Statistical Expertise, focusing on consulting projects related to the environment.

Anthony R. Olsen is an environmental statistician with the Western Ecology Division, National Health and Environmental Effects Research Laboratory, U.S. Environmental Protection Agency. He received his Ph.D. in 1973 from the Oregon State University Department of Statistics and his B.S. in 1966 and M.S. in 1969 from the University of Wyoming Department of Statistics. His research interests include the design of large-scale environmental monitoring programs, statistical graphics, and environmental statistics. Previously, Dr. Olsen was a senior research statistician and statistics group manager at Battelle Pacific Northwest Laboratories, where he conducted studies related to atmospheric modeling, acid deposition, and other environmental issues.

CHAPTER 2

CHEMICAL ASSAY VALIDATION

Russell Reeve and Francis Giesbrecht

Many manufacturing processes depend upon measurements made on the product of the process. To maintain control over the manufacturing process, these measurements must themselves come from a measuring process of satisfactory quality. Therefore, an assessment of the characteristics of the measurement process is important. This case study discusses the statistical analysis of a measuring process set in the pharmaceutical industry: assay validation. Here we discuss one facet of assay validation: the assay's accuracy and repeatability.

While the terminology of this case study comes out of the pharmaceutical/biotechnology industries, the statistical reasoning crosses the boundaries of many industries.

INTRODUCTION

In the pharmaceutical industry, chemical assays must be validated before use in pharmacokinetic[1]/pharmacodynamic[2] studies, manufacturing, or stability[3] analyses. A method validation is very similar in principle to gage studies (defined below) found in other industries; however, a validation is more extensive. We will discuss one component of a validation package: the analysis of a method's accuracy and precision. In general, the accuracy refers to the bias of a method, while precision refers to the variability in a method, usually measured by the coefficient of variation (CV); in some laboratories, the CV is called the relative standard deviation, or RSD for short.

BACKGROUND INFORMATION

A gage study is any study of a measuring process designed to assess the measuring process' capability. The chief concerns in a gage study are the measuring process' accuracy, reproducibility, and repeatability. Repeatability is the standard deviation (or,

---

[1] This is a study designed to estimate the distribution and absorption of a drug in the body.
[2] This is a study designed to estimate the effect of a drug on the pharmacological processes of a body.
[3] The ability of a drug to remain potent, undegraded over time.

equivalently, CV) within a single measurement run (i.e., one operator on one day on one piece of equipment, etc.); reproducibility is the standard deviation (or CV) when all sources of variation are accounted for, including operators, days, etc. The term is derived from a gage block, which is a block used to measure small distances. The analysis for accuracy and precision (repeatability and reproducibility) in the assay validation is the same as that for a gage study. In this case study, we will be discussing the validation of a High Performance Liquid Chromatography (HPLC) method for the analysis of the potency of a drug product. Every batch of drug product must be analyzed for potency before it may be released for use. A batch with potency above specification could prove to have side effects or toxicity; a batch with potency below specification could fail to be efficacious. HPLC methods are also used for stability studies, for dissolution[4] studies, and occasionally for bioavailability[5] and pharmacokinetic studies.

In an HPLC assay, the compound of interest is measured by first separating it from other compounds likely to be found with it and then measuring how many molecules are in the sample by the amount of light absorbance at some frequency; see Fig. 1. The drug product is dissolved in a solution; this solution is then injected into a column (Point A). The column is filled with a small pellet powder,[6] known as the stationary phase. As the solution is pushed through this powder at high pressure (often more than 1500 psi), the smaller molecules will tend to come out of the column (Point B) faster than the larger molecules, thus separating the drug molecules from the other molecules found with it. However, other factors, such as pH, will also affect the order in which the molecules will exit the column, due to factors such as electrical interaction between the molecules in the mobile phase with those in the stationary phase, etc. The time at which the drug molecules exit the column is known. The solution is passed in front of a detector (Point C), where a light beam of a specific wavelength is passed across the solution. The detector produces a voltage proportional to the amount of light reaching it. If the light is set to an appropriate frequency, the quantity of light passing through the solution is proportional to the number of molecules of the drug between the lamp and the detector.

Figure 2 has a picture of a typical response from a detector. The area under the peak found at the time the drug product is known to be found is calculated, and that is the response used for the calibration.

A calibration curve is used to estimate the quantity of drug in a sample. See Fig. 3 for an example of a calibration curve. In an HPLC run, several samples of known drug amount are analyzed; these are called the standards, and they are often expressed as amounts added. The peak areas of the standards are regressed onto the drug amounts, producing the calibration curve; for HPLC assays, the calibration curve is typically linear. The peak areas of the unknown samples are then interpolated off this calibration curve, yielding estimates of their true drug amount, often called amounts found. These estimates are typically expressed in terms of percentage of label strength, or %LS for short.

---

[4] How fast a solid dosage or lyophilized drug dissolves.
[5] How much and for how long a drug is in the system, available to act (bioavailable).
[6] The HPLC's solid phase's composition will change depending on the properties of the analyte or the other compounds in the mixture. The analyte is that compound whose content we are analyzing. The mixture in which all the compounds live is called the matrix.

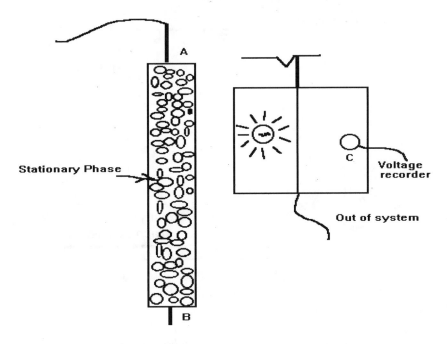

Fig. 1. *Schematic diagram of a basic HPLC system.*

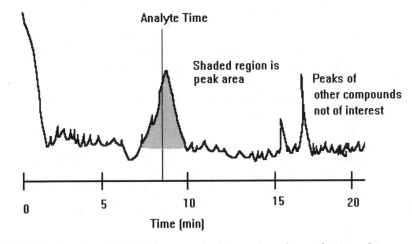

Fig. 2. *Signal produced by HPLC system; shaded area is peak area that is used to compute the %LS Found using the calibration curve. The time is elapsed time since sample was injected into Point A.*

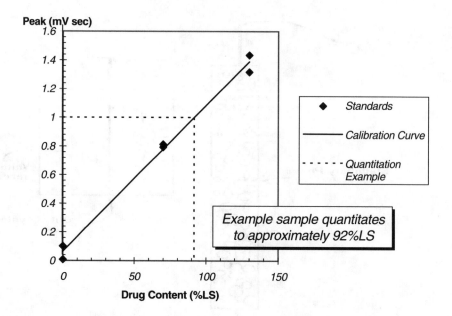

Fig. 3. *Example of a standard curve and a quantitation of a sample that yielded a peak area of 1 using the curve. Noise in the standards has been exaggerated to make it easier to see.*

QUESTION(S) OF INTEREST

To validate the method, several samples of known analyte content are prepared (%LS Added) at several concentrations (usually 3 to 5) and analyzed, with the method reporting back an estimated content (the %LS Found). In an ideal assay, the estimated content would equal the known content.

The questions the chemist wishes to ask are as follows:
1. What is the method's accuracy? Is it unbiased?
2. Is the method precise? A CV ≤ 2% is acceptable; over 2% is unacceptable.
    Note that the CV above refers to repeatability.

In a complete validation, other factors would also need to be investigated: reproducibility, specificity of the method, robustness to small changes in procedure or equipment settings, system suitability, etc. (e.g., [Chow and Lui, 1995], [FDA, 1987], [Shah et al., 1992]). We will not consider these other issues here.

DATA

Name of Data File: Case02.txt
Format: Tab delimited

Variable Name   Description
%LS Added   Amount of analyte spiked into the sample solution to be analyzed
%Recovery   Amount of analyte quantitated divided by %LS Added;
                      expressed as a percentage.

Table 1. *Recovery data for an HPLC method collected to investigate the method's accuracy. The data are derived from ten independent standards.*

%LS Added	%Recovery
70	99.72
70	100.88
85	100.82
85	100.50
100	100.47
100	101.05
115	101.85
115	101.44
130	101.22
130	101.93

These accuracy data are for a drug to treat organ rejection after transplants; the dosage form is a tablet. The %Recovery was found by taking the %LS Found and dividing it by the %LS Added, and then multiplying by 100%. %Recovery is preferred to %LS Found since it is interpretable without regard to the %LS Added. Since we do not know which %LS Added we will have in actual samples, this generality is quite useful. Also, the standard deviation tends to be proportional to the amount added; hence, %Recovery is homoscedastic. These data were obtained on a single HPLC run where standards were quantitated (peak areas not given as they are not needed for this case study).

ANALYSIS

The regression model considered for these data is %Recovery = $\beta_0 + \beta_1$ (%LS Added) + ε. The closer β_0 is to 100 and β_1 is to 0, the better in terms of assay performance. It can be safely assumed that the errors ε are approximately normally distributed on this scale, with unknown constant variance σ^2.

There are several approaches that one can take to validate a method for accuracy and precision. We will present three, in order of commonness. Your instructor will inform you which approach(es) to pursue.

APPROACH 1: Hypothesis Testing Approach

We will accept the method if we do not reject the null hypothesis H_0 (either version (a) or (b)). Test either (a) or (b) at the 5% level.

(a) Joint hypothesis
H_0: $\beta_0 = 100$, $\beta_1 = 0$, $\sigma \leq 2$
vs.
H_A: not H_0

(b) Three single hypotheses
H_0: $\beta_0 = 100$ vs. H_A: $\beta_0 \neq 100$
H_0: $\beta_1 = 0$ vs. H_A: $\beta_1 \neq 0$
H_0: $\sigma \leq 2$ vs. H_A: $\sigma > 2$

1. What are your conclusions as to the method acceptability?
2. What is the power of the test?
3. Compute confidence intervals β_0, β_1, and σ.

APPROACH 2: Confidence Interval Approach

We will accept the method if the confidence interval for the mean predicted value is within ±2 %LS over the range of the data and if $s \leq 2$ %LS. Note that s is the point estimate of σ.

1. Is the method acceptable under these conditions?

APPROACH 3: Tolerance Interval Approach

We will accept the method if the 95% tolerance interval for 95% of the center of the population is within ±6% over the range of the data (cf. [Patel, 1986]).

1. Is the method acceptable under these conditions?

INSTRUCTIONS FOR PRESENTATION OF RESULTS

The results should be presented in a written report suitable for presenting to an audience with a wide variation in statistical expertise. The parts of the report should include the following:

Report Section	Statistical expertise expected
Management Summary	Very little; also has very little time
Conclusion	Not interested in the statistics; plots very useful
Statistical Methodology	To be reviewed by statistical peers (possibly at FDA or internally)
Data	Present the data so as to avoid ambiguity later

REFERENCES

Blackwelder, W. C. (1982), "Proving the Null Hypothesis" in Clinical Trials, *Controlled Clinical Trials*, 3, pp. 345–353.

Chow, Shein-Chung and Jen-Pei Lui (1995), *Statistical Design and Analysis in Pharmaceutical Science*, Marcel Dekker, New York, Chapter 3.

FDA (1987), *Guideline for Submitting Samples and Analytical Data for Methods Validation.*

Patel, J.K. (1986), Tolerance limits—A review, *Commun. Statist.-Theor. Meth.*, 15(9), pp. 2719–2762.

Shah, Vinod P., et al., (1992), Analytical methods validation: Bioavailability, bioequivalence, and pharmacokinetic studies, *Pharmaceutical Research*, V. 9.4, pp. 588–592.

Schuirmann, D. J. (1987), A comparison of the two one-sided tests procedure and the power approach for assessing the equivalence of average bioavailability, *J. Pharmacokinet. Biopharm.*, 16(6), pp. 657–680.

BIOGRAPHIES

Russell Reeve attended Utah State University, where he received a B.S. in Mathematics. It was there that he was introduced to statistics by Prof. Donald Sisson. After obtaining an M.S. in Operations Research and Statistics from Rensselaer Polytechnic Institute and a Ph.D. in Statistics from Virginia Polytechnic Institute and State University, he had brief stints teaching statistics at the University of Rochester and Memphis State University. He found a position with Syntex Research in 1992 providing statistical expertise and guidance to analytical chemists, and later to drug formulators and to biologists. He now works at

Chiron providing statistical support for assay development and validation for vaccines, therapeutics, and diagnostics.

Francis Giesbrecht is Professor of Statistics at North Carolina State University. He received a Bachelor of Science in Agriculture degree from the University of Manitoba in Canada, an M.Sc. degree in Agriculture from Iowa State University, and a Ph.D. in Statistics from Iowa State University. He has been on the faculty at NC State University since 1971. He has an appointment in both the College of Physical and Mathematical Sciences and in the College of Agriculture and Life Sciences. He has taught a range of courses, including Mathematical Statistics, Statistical Methods for Biological Sciences, Social Sciences, and, for Engineers, Design of Experiments and Linear Models and Variance Components. He spends a significant portion of his time serving as a statistical consultant on projects in many departments on campus.

CHAPTER 3

AUTOMATING A MANUAL TELEPHONE PROCESS

Mary Batcher, Kevin Cecco, and Dennis Lin

This article was written and prepared by U.S. Government employees on official time. It is in the public domain and not subject to U.S. copyright. The content of this article is the opinion of the writer and does not necessarily represent the position of the Internal Revenue Service. The mention of specific product or service in this article does not imply endorsement by any agency of the federal government to the exclusion of others which may be suitable.

This case study illustrates the use of statistics to evaluate a new technology that will be implemented nationally in over 30 locations if it proves successful in a pilot study. Specifically, the case study is of an interactive telephone application that will let certain types of calls to the IRS be handled without the intervention of a staff person.

When introducing new technology, it is important to consider the human interaction with that technology. The technology may function perfectly but the ability or willingness of people to use it may not be there. It is thus very important to pilot test new systems. The interactive telephone system was pilot tested and evaluated in terms of cost, customer satisfaction, and ease of use. This case study focuses on the assessment of cost, in terms of time to complete a transaction, and ease of use, in terms of the percent of users who successfully completed their transaction without requiring the assistance of IRS staff.[1] The case study illustrates the use of hypothesis testing in decision making and the use of basic time series statistics and plots to examine periodic fluctuation over time.

---

[1] The customer satisfaction measure was an automated survey of a sample of callers. The caller was asked, at the end of the transaction, to participate in a short survey. The general satisfaction level from the survey was 93.6 percent and the survey response rate was 64.4 percent.

Introduction

The problem to be solved is to assess the effectiveness of a new interactive telephone system by examining its performance in a pilot test. The goals of the statistical analysis are twofold. The first goal is to determine whether a new interactive telephone system is more efficient than the manual process it replaces, as measured by the time needed to complete a transaction. The second is to examine ease of use and the effect of extraneous occurrences in the pilot test by tracking successful completion rates over time to identify any problematic occurrences or any periodicity in the data.

Background Information

Interactive telephone systems allow us to exchange limited information using the push buttons on our touch-tone telephones. The most basic of these systems allows us to route ourselves to a recorded message or to the best person to handle our issue. More complex systems allow a complete exchange of information, with problems fully resolved in the automated setting. Interactive telephone systems are increasingly common. We use them to check bank balances, order merchandise, inquire about service, etc. They offer opportunities to decrease costs while maintaining or increasing accessibility. The IRS is developing interactive systems for some of their telephone work. When people are unable to pay all of the taxes they owe, they are able to establish installment agreements with the IRS to pay off the balance over time. One of the interactive telephone systems developed by the IRS allows people to request a payment extension or establish an installment agreement. (Additional examples of the use of telephone pushbuttons for the automated exchange of information can be found in [Nicholls & Appel, 1994], [Rosen, Clayton, and Pivetz, 1994], and [Werking, Clayton, and Harrell, 1996].)

The IRS has developed the Voice Balance Due (VBD) fully automated telephone system. It permits people who call the IRS for the purpose of requesting a payment extension or establishing a monthly payment plan to do so by using the push buttons on their telephone. The system is fully automated for eligible callers and does not require any contact with an IRS representative. The VBD system automates a process that is currently manual.

During the pilot test of the VBD system, taxpayers receiving IRS balance due notices were also notified that they might be eligible to use their touch-tone telephone to request an extension of time to pay their taxes or to set up a schedule to pay their taxes in installments. Callers to the VBD system had to enter their social security number and a unique caller identification number printed on the notice. There were checks to screen out ineligible callers based on several criteria. Once into the system, the caller was given the option to select a payment extension or a monthly payment plan. Those who selected a payment plan were asked to enter the amount they could afford to pay each month and the day of the month they wished to establish as their monthly due date. Callers who established a monthly installment plan or a payment extension received a confirmation letter from the IRS. The confirmation letter included the terms and conditions of the arrangement.

Some of the anticipated advantages to the VBD system are that it can provide callers better access to the IRS in terms of increased capacity to handle calls. Twenty-four–hour access was not available because throughout the call, callers were given the option of exiting the automated system to speak with an IRS representative. The establishment of fully automated telephone systems also provides the IRS the opportunity to redirect staff to deal with more complex issues. A goal for the VBD system is to provide the same or better service to callers with greater efficiency than the manual process.

QUESTIONS OF INTEREST

New systems and procedures must be evaluated not only in terms of their cost savings or operating efficiencies but also in terms of their ability to provide service that is at least as easy to use as the system they are replacing. The automated installment agreement application of the VBD system replaces a manual process for the less complex installment agreements. One of the ways that we can measure ease of use is by examining the percentage of callers who succeed in establishing installment agreements through the VBD system out of those eligible callers who attempt to set up agreements. Callers can default out of the automated process at any time to an IRS employee. Therefore, the percent of callers who successfully complete an automated agreement is an indicator that the system is easy for them to use. Ease of use is affected by many factors, including the clarity of the information mailed to taxpayers informing them that they owe a balance and inviting them to use the VBD system, the clarity of the telephone script and on-line prompts, and the motivation and general capability of the caller. We might speculate that there are differences in the ability of callers to use the VBD system related to day of the week, time of day, etc. During the pilot test, there was some variability in the material mailed to taxpayers; this might also account for some differences in ease of use of the system. Beginning in June 1995, material was added to the mailing to taxpayers informing them that they might be eligible to use the VBD system to establish an installment agreement and providing some information about using the system. This additional material was discontinued in December. There is also some clustering of the mailings of notices.

To assess ease of use, we can explore the following questions: To what extent have callers been served by the VBD application; i.e., what percent of eligible callers successfully completed the application, and did that percent differ over the period of the pilot test? Were there any periodic fluctuations in completion rates?

Major goals of automated systems are to increase access and decrease costs. Cost can be measured in terms of efficiency. Automated systems are valuable primarily to the extent that they handle calls more efficiently than manual processes.

To assess system efficiency we would like to know the following: Is it more efficient for the IRS to have taxpayers set up installment agreements manually or through the use of the VBD automated system?

Note: A limited study in a single district found that, in 60 calls handled manually, the average time needed to complete an installment agreement was 15 minutes and 38 seconds. The standard error associated with this estimate of mean time was 4 minutes and 58 seconds. This is our best information about the time needed to complete a manual installment agreement of the type handled by the VBD system.

DATA

Many more variables are captured by the VBD system than are presented in this case study. We selected a subset of variables that have direct bearing on the questions posed above. They include the date (Monday through Friday only), daily counts of the number of callers requesting an installment agreement, the number of callers successfully completing an installment agreement, and the day's average call length for a completed agreement.

Data collected by the automated telephone system are free from many of the measurement issues that exist in data collected manually. However, there are still issues about the refinement of the measure. Our data are presented as daily counts rather than as specific information about each call. This precludes us from tracing the success of particular types of calls as they progress through the system and limits the types of analysis that can be done.

Although the pilot test period was from April 24 through June 23, 1995, we have included additional data through December 8, 1995.

Variable Name   Description

Date   mm/dd/yy
Request   Callers Requesting an Installment Agreement Plan
Complete   Callers Successfully Completing an Installment Agreement
Length   Average Call Length, in seconds, for a Completed Payment Agreement

Date	Request	Complete	Length
042495	25	22	206
042595	15	11	172
.	.	.	.
.	.	.	.
.	.	.	.
120895	41	34	194

ANALYSIS

Two questions were identified in Questions of Interest as the key issues to be addressed in the statistical analysis. They are listed below with suggested analytic approaches.

Question 1

To what extent have callers been served by the VBD application; i.e., what percent of eligible callers successfully completed the application, and did that percent differ over the period of the pilot test? Were there any periodic fluctuations in completion rates?

Question 1 can be answered by the following statistical techniques: (1) Convert raw data into a meaningful picture of completion rates on a daily and weekly basis. (2) Create graphs of the daily and weekly rates and examine the data over time. (See Figures 1 and 2.) (3) Calculate and examine the autocorrelates to determine whether or not there appears to be any significant fluctuation in time (e.g., from week to week or day to day) among the completion rates. Statistical methods we recommend are time series plot for both weekly and daily data, as well as autocorrelation plots. (See Figures 3 and 4.) (Useful time series references include [Box, Jenkins, and Reinsel, 1994] and [Wei, 1990].) The Pearson correlation coefficient is probably the most common measure for the dependency between two random variables. When there is only one random variable in a time sequence, we can evaluate the autocorrelation to measure the dependency among all the observations along the time sequence.

For example, given a time series $y = (y_t, y_{(t-1)}, ..., y_2, y_1)$, we can artificially generate a lag $- 1$ series $y_{-1} = (y_{(t-1)}, y_{(t-2)}, ..., y_2, y_1)$ and then evaluate the correlation between the first $(t-1)$ observations from y and y_{-1}. This is called the autocorrelation of lag -1. If such an autocorrelation is significant, we know that the observations are not independent, and in fact, each observation is correlated to the previous observation. Similarly, we can define autocorrelations of lag –2, lag –3, and so on to investigate the dependency among all observations.

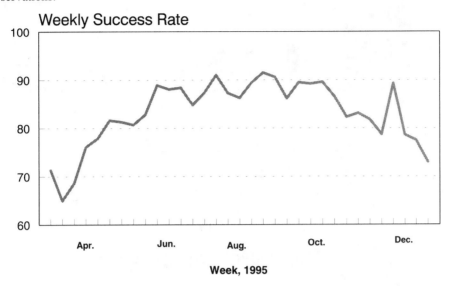

Fig. 1. *Weekly percentage of successfully completed installment agreements.*

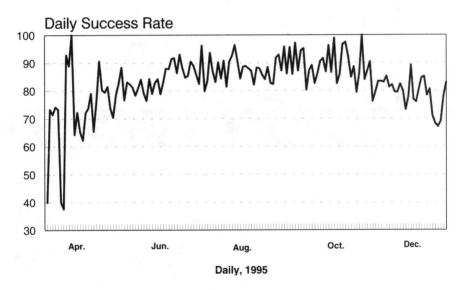

Fig. 2. *Daily percentage of successfully completed installment agreements.*

Figure 3 (autocorrelates for lags 1 through 10 for VBD data summarized on a weekly basis) shows a significant fluctuation in the success rates during the first two weeks of the pilot. This would be indicative of correcting unexpected problems that quite frequently occur in the earliest stages of a pilot test. The remaining eight weeks (lags) shown on Figure 3 reveal no significant change from one week to the next in the completion rates.

Batcher, Cecco, and Lin

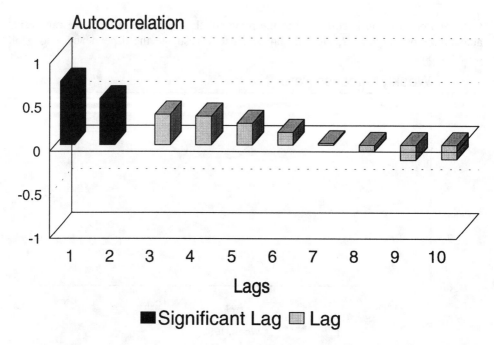

Fig. 3. *Autocorrelations for lags 1 through 10 for VBD data summarized on a weekly basis.*

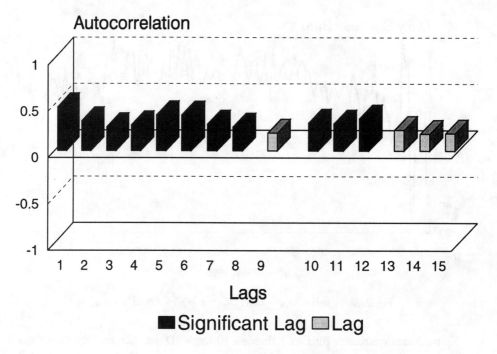

Fig. 4. *Autocorrelations for lags 1 through 15 for VBD data summarized on a daily basis.*

Question 2

Is it more efficient for the IRS to have taxpayers set up installment agreements manually or through the use of the VBD automated system?

Question 2 can be answered by the following statistical techniques: (1) Compute the average automated time to complete an automated installment agreement and determine the variability (standard error) associated with the estimate. (2) Conduct a hypothesis test where the null hypothesis is that the average length of time to complete a manual installment agreement is the same as the length of time to complete an automated installment agreement and the alternative hypothesis is that the average length of time to complete a manual installment agreement is greater than the length of time to complete an automated installment agreement. Namely, test the hypothesis Ho: $\mu_{Lm} = \mu_{Li}$ vs. Ha: $\mu_{Lm} > \mu_{Li}$, where $\mu_{Lm} =$ average time to complete a manual installment agreement and $\mu_{Li} =$ average length of time to complete an automated installment agreement. Evaluate the decision with some level of confidence. Statistical methods we recommend here are descriptive statistics, hypothesis testing, two-sample t test, and the F test to test the equality of variance assumption. We have found [Berenson & Levine, 1993], [Keller and Warrack, 1997], [Siegel, 1994], and [Vardeman, 1994] to be useful references for introductory statistical methods.

INSTRUCTIONS FOR PRESENTATION OF RESULTS

For Question 1, results should be presented with a brief written description of the findings, accompanied by graphic displays. Some statement about the apparent ease of use should be included, as well as a recommendation for any additional data collection and analysis.

The "trend patterns" can easily be seen from the time series plot for weekly data, while the "seasonal patterns" are most easily shown in the time series plot of the daily data. The trend pattern can be captured by the simple linear regression, and the seasonal pattern can be evaluated by the autocorrelation plot (as shown in Figures 3 and 4). In contrast to most introductory statistical content, the "independent" assumption is in general not valid in time series data.

For Question 2, results should be summarized in a written report, accompanied by a recommendation about the implementation of the VBD system, based on system efficiency.

Here we have used some descriptive statistics to illustrate the basic comparison. A statistical test of the hypotheses Ho: $\mu_{Lm} = \mu_{Li}$ versus Ha: $\mu_{Lm} > \mu_{Li}$ will be sufficient to answer Question 2. We also suggest the simple comparison based on confidence intervals. This seems to be an easier approach to helping undergraduate students grasp the idea of making comparisons in the presence of uncertainty.

A report should be generated that addresses the findings. It should include graphics and a discussion of the limitations of the study and recommendations for further analysis. It should also include a recommendation about adopting the VBD system based on system efficiency and ease of use. There should be a discussion about the effectiveness of these measures for their stated purpose.

REFERENCES

Berenson, Mark L. and Levine, David M. (1993), *Statistics for Business and Economics*, Prentice–Hall: Englewood Cliffs.

Box, G.E.P., Jenkins, G.M., and Reinsel, G.C. (1994), *Time Series Analysis Forecasting and Control, Third Edition*, Prentice–Hall: Englewood Cliffs.

Keller, Gerald and Warrack, Brian (1997), *Statistics for Management and Economics*, Duxbury Press: Belmont.

Nicholls II, William L. and Appel, Martin V. (1994), "New CASIC Technologies at the U.S. Bureau of the Census," *Proceedings of the Section on Survey Research Methods, American Statistical Association*, pp. 757–762.

Rosen, Richard J., Clayton, Richard L., & Pivetz, Lynda L. (1994), "Converting Mail Reporters to Touchtone Data Entry," *Proceedings of the Section on Survey Research Methods, American Statistical Association*, pp. 763–768.

Siegel, Andrew F. (1994), *Practical Business Statistics*, Irwin: Boston.

Vardeman, Stephan B. (1994), *Statistics for Engineering Problem Solving*, PWS Publishing Company: Boston.

Wei, William W.S. (1990), *Time Series Analysis*, Addison–Wesley: Redwood City.

Werking, George S., Clayton, Richard L., and Harrell Jr., Louis J. (1996), "TDE and Beyond: Feasibility Test of E-Mail/World Wide Web for Survey Data Collection," *Proceedings of the Section on Survey Research Methods, American Statistical Association*.

BIOGRAPHIES

Mary Batcher manages a statistical consulting group at the Internal Revenue Service (IRS). The group consults on a number of internal projects such as testing new systems, designing surveys, and designing samples. Before coming to the IRS, she was at the National Center for Education Statistics.

Kevin Cecco is a consulting statistician at the IRS, working on the development of new automated systems for handling some telephone calls and for the submission of simple tax returns. He came to the IRS from the Bureau of the Census.

Dennis Lin is an associate professor at Penn State University, where he holds a joint appointment in the Departments of Statistics and Management Science. Prior to his appointment at Penn State, he was an associate professor at the University of Tennessee.

CHAPTER 4

DISSOLUTION METHOD EQUIVALENCE

Russell Reeve and Francis Giesbrecht

In this case study, we will explore the concept of equivalence, and in particular how it relates to a dissolution test. Many questions that are answered with hypothesis testing could be better answered using an equivalence approach. A discussion of dissolution tests will come first; it will be followed by an explanation of why we need criteria for equivalence. It is helpful if the reader has at least read the Chemical Assay Validation case study (Chapter 2) before proceeding with this case study, though it is not necessary.

INTRODUCTION

There are two issues in this case study: (1) Devise statistical criteria to decide if two sites yield equivalent results, and (2) apply the methodology developed on two data sets. Note that (1) is more theoretical in nature, while (2) applies the mathematical work of (1) to data. Since many comparisons are best dealt with as equivalence problems, the methodology has wide applicability.

BACKGROUND INFORMATION

A dissolution test measures how fast a solid-dosage pharmaceutical product dissolves [USP XXII], [Cohen et al., 1990]. Since variation in dissolution profiles can have deleterious effects on the *in vivo*[1] performance of a solid-dosage product, a test that measures the dissolution is of upmost importance.

The dissolution apparatus often consists of six vessels, each containing a dissolving solution—typically water with pH adjusted to cause the tablets or capsules to dissolve, though in some cases digestive enzymes may be used to more realistically simulate that action of the stomach. The sampled units are dropped into the vessels; the units here are either individual capsules or tablets; see Fig. 1. The vessels themselves are in a water bath to maintain a nearly constant temperature. The solution containing the capsule or tablet is

---

[1] *In vivo* means in either human or animals; *in vitro* means in the laboratory (e.g., in the test tube).

stirred with a paddle rotating at a fixed rotational velocity. At predetermined times, the probe withdraws a small amount of the solution, and that sample is then analyzed, typically by an HPLC[2] system. The assayed values are then expressed on a percent of label strength (%LS) or percent dissolved basis, and the plots of the amount dissolved (released) as a function of time are generated; these plots represent the dissolution profile. See Fig. 2 for a typical profile.

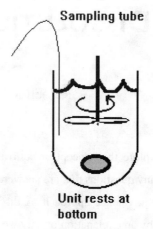

Fig. 1. *Schematic of physical setup of dissolution apparatus.*

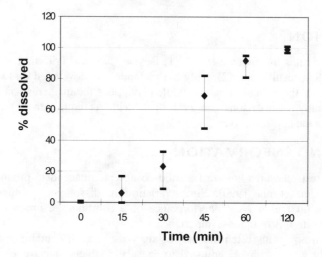

Fig. 2. *Typical dissolution profile.*

---

[2] High Performance Liquid Chromatography; the device quantitates how much of the drug is in the solution fed to it, with some amount of random variation. The exact nature of the method is not important for understanding this case study or for understanding dissolution tests.

The dissolution is often developed in one laboratory, typically the company's research and development center, and then transferred to another site for use, typically the manufacturing facility located at another site; however, transfers even within the R&D site are not uncommon. Because of federal regulations described in Good Manufacturing Practices for Finished Pharmaceuticals [21 CFR 211.160], before the new site can use the method, it must be verified that the new site can produce results equivalent to those of the originating site. Note that one should treat the concept of "site" liberally here, as even a transfer to another analyst within the same site requires documentation that the new analyst can yield "equivalent" results to the originating analyst. Therefore, some concept of equivalence is needed.

Two different dissolution equivalence problems arise: that for immediate-release products, and that for sustained-release products. An immediate-release product is one that is designed to dissolve and enter the bloodstream as fast as possible. In this case, one typically wants an assurance that the product is at least $p\%$ dissolved, for some p (between 0 and 100), at a specified time point (possibly more than one time point may be specified). A sustained-release product is one that has been designed to slowly dissolve so that there is a more constant concentration of drug in the body. Here, one wants some assurance that the product is dissolving at a controlled rate, neither too fast nor too slow. Typically, the percent dissolved must pass through prespecified hoops at several time points (3 not being uncommon).

QUESTION(S) OF INTEREST

To keep the problem from becoming extremely difficult, each time point will be considered separately; i.e., we want to construct a univariate procedure rather than a joint procedure. In practice, one would prefer the joint procedure since the data are correlated within a vessel. Since the variation in the %LS values vary as a function of the mean, different criteria will be needed for different mean values. In general, the most variation is found around 70 %LS, with the least around 0 %LS and 100 %LS. All units have potencies that are close to 100 %LS; the unit-to-unit variation is, in general, considerably less than the method variation. For this problem, criteria will be developed for 2 time points; which time points will be explained in the next paragraph. The accepted facts are listed in Table 1.

The products we are considering in this case study are both immediate-release products. Therefore, we need equivalence criteria for two points: (a) The time point where the %dissolved is closest to 70 but definitely less than 90, and (b) the last time point. The time point as described in (a) is chosen since that is often the most discriminating time point (i.e., can differentiate two methods, two lots of product, etc., the most easily, despite also having the largest standard deviation). The time point as described in (b) is chosen since that is the time at which the drug product should be completely dissolved.

From long-standing experience, it can be safely assumed that the within-site variances are common across sites.

After the criteria for calling two dissolution profiles equivalent has been constructed, two profiles must be compared for equivalence. Both products are scaling up production after regulatory approval to market these products. Hence, they are being moved to larger facilities.

Table 1. *Accepted bounds for dissolution equivalence.*

If mean dissolved is	Mean differences of this much are considered scientifically important	An approximate upper bound for the standard deviation* is
< 90 %LS	± 15 %LS	8
≥ 90 %LS	± 7 %LS	3

*Average across products of the maximal standard deviation with mean %dissolved in the listed category

DATA

Name of Data File:	Case04.txt
Format:	Tab delimited

Variable Name	Description
Drug	Drug of comparison; note that each drug should be treated individually
Site	Site of laboratory performing test
Time	Time (in minutes) of sampling of dissolution solution
Vessel1	%Dissolved in vessel 1
Vessel2	%Dissolved in vessel 2
etc.	

The first five rows of the data listing are shown in Table 2. In Tables 3 and 4, the complete data listing for each drug is presented. Table 3 lists the dissolution data for a growth hormone, fed to cattle as a feed additive to help add muscle mass. It is formulated as a capsule. The dissolution media has a pH of 4.0; paddle speed is 50 rpm, and the temperature is 30 °C. Table 4 lists the dissolution data for an analgesic in tablet form; it is water soluble. The dissolution media is deionized water, paddle speed is 40 rpm, and the temperature is 25 °C.

Table 2. *First five observations of the data file "Case04.txt."*

Drug	Site	Time	Vessel 1	Vessel 2	Vessel 3	Vessel 4	Vessel 5	Vessel 6
FeedAdditive	New York	0	0	0	0	1	0	0
FeedAdditive	New York	15	2	5	0	17	1	12
FeedAdditive	New York	30	20	33	9	23	23	32
FeedAdditive	New York	45	65	82	48	81	77	61
FeedAdditive	New York	60	95	92	81	94	95	93

Table 3. *Drug = cattle feed additive. Compare New York vs. California.*

		%Dissolved					
Site	Time (min)	Vessel 1	Vessel 2	Vessel 3	Vessel 4	Vessel 5	Vessel 6
New York	0	0	0	0	1	0	0
	15	2	5	0	17	1	12
	30	20	33	9	23	23	32
	45	65	82	48	81	77	61
	60	95	92	81	94	95	93
	120	98	97	101	100	99	99
California	0	0	0	0	0	0	0
	15	13	25	27	2	17	18
	30	28	42	45	8	28	38
	45	35	55	70	50	41	63
	60	90	99	95	85	78	96
	120	99	103	100	97	100	101

Table 4. *Drug = Analgesic. Compare New Jersey to Puerto Rico.*

		%Dissolved					
Site	Time (min)	Vessel 1	Vessel 2	Vessel 3	Vessel 4	Vessel 5	Vessel 6
New Jersey	0	0	0	0	0	0	0
	20	5	10	2	7	6	0
	40	72	79	81	70	72	73
	60	96	99	93	95	96	99
	120	99	99	96	100	98	100
Puerto Rico	0	0	0	0	0	0	0
	20	10	12	7	3	5	14
	40	65	66	71	70	74	69
	60	95	99	98	94	90	92
	120	100	102	98	99	97	100

ANALYSIS

A. Theoretical Problem

1. Devise a simple, yet statistically valid, rule for deciding whether two sites have equivalent results. For generality, the rule should depend on the sample size. Sample sizes that are possible are 6, 12, 18, and 24. Sample sizes larger than 24 would put too onerous a burden on the quality control laboratory.
2. Construct an operating characteristic (OC) curve of the rule as a function of mean difference. Note that $OC(\Delta) = \text{Prob}\{\text{declare profile pairs not equivalent} \mid \mu_1 - \mu_2 = \Delta\}$.
3. What sample size should be used? Again, the sample sizes possible are 6, 12, 18, and 24. The restrictions to multiples of 6 are because of the design of the testing apparatus. If the true mean difference is within 1 standard deviation, then we would like to accept the methods as being equivalent at least 80% of the time.

B. Application

Apply the rule devised in 1 above to the two data sets in Tables 3 and 4.

INSTRUCTIONS FOR PRESENTATION OF RESULTS

The results should be presented in a written report suitable for presenting to an audience with a wide variation in statistical expertise. The parts of the report should include the following:

Report Section	Statistical Expertise Expected
Management Summary	Very little; also has very little time
Conclusion	Not interested in the statistics; plots very useful
Statistical Methodology	To be reviewed by statistical peers (possibly at FDA or internally)
Data	Present the data so as to avoid ambiguity later

REFERENCES

21CFR211.160, Code of Federal Regulations, Title 21, Vol. 4, Parts 200–299, Revised April 1, 1997.

Cohen, J. L. et al. (1990), The development of USP dissolution and drug release standards, *Pharmaceutical Research*, Vol. 7, No. 10, pp. 983–987.

United States Pharmacopeia xxii, USP-NF xxii, p. 1579.

BIOGRAPHIES

Russell Reeve attended Utah State University, where he received a B.S. in Mathematics. It was there that he was introduced to statistics by Prof. Donald Sisson. After obtaining an M.S. in Operations Research and Statistics from Rensselaer Polytechnic Institute and a Ph.D. in Statistics from Virginia Polytechnic Institute and State University, he had brief stints teaching statistics at the University of Rochester and Memphis State University. He joined Syntex Research in 1992, providing statistical expertise and guidance to analytical chemists, and later to drug formulators and biologists. He now works at Chiron providing statistical support for assay development and validation for vaccines, therapeutics, and diagnostics.

Francis Giesbrecht is a Professor of Statistics at North Carolina State University. He received a Bachelor of Science in Agriculture degree from the University of Manitoba in Canada, an M.Sc. degree in Agriculture from Iowa State University, and a Ph.D. in Statistics from Iowa State University. He has been on the faculty at NC State University since 1971. He has an appointment in both the College of Physical and Mathematical Sciences and in the College of Agriculture and Life Sciences. He has taught a range of courses, including Mathematical Statistics, Statistical Methods for Biological Sciences, Social Sciences, and, for Engineers, Design of Experiments and Linear Models and Variance Components. He spends a significant portion of his time serving as a Statistical consultant on projects in many departments on campus.

CHAPTER 5

COMPARISON OF HOSPITAL LENGTH OF STAY BETWEEN TWO INSURERS FOR PATIENTS WITH PEDIATRIC ASTHMA

Robert L. Houchens and Nancy Schoeps

This case study investigates the relative importance of several factors in predicting the length of time young patients with asthma stay in the hospital. With the present atmosphere of cutting health care costs it is important to look at providing adequate care while at the same time reducing costs and not keeping children in the hospital longer than necessary. By looking at a sample of patients with pediatric asthma, concomitant factors to the main reason for being admitted to the hospital may shed light on different lengths of stay.

INTRODUCTION

In today's healthcare environment, health insurance companies are increasingly pressuring hospitals to provide high quality health services at the lowest possible cost. The vast majority of all healthcare costs are for hospitalization. During the past decade, inpatient costs of patients in hospitals have been reduced in two primary ways. First, the less severe cases are now treated in the doctor's office or in hospital emergency rooms rather than being admitted to the hospital. Second, for cases admitted to the hospital, the lengths of hospital stays have been considerably shortened.

It is believed that some insurers have been more successful than others at minimizing hospital lengths of stay (LOS). To test this, a sample of hospital medical records was drawn for each of several illnesses from metropolitan hospitals operating in one state. The data for this case study consists of information abstracted from the medical records of asthma patients between the ages of 2 and 18 years old.

BACKGROUND

The sample of medical records was drawn in two stages. At the first stage, 29 metropolitan hospitals were sampled with probabilities proportional to an estimate of the number of asthma admissions it had during a single year. At the second stage, 393 asthma cases insured by Insurer A were randomly selected and 396 asthma cases insured by Insurer B were randomly selected from the 29 hospitals.

Information was abstracted from each patient's medical record. Aside from the main variables of interest, insurer and LOS, the additional information falls into four categories:

A. Patient Severity Variables describe the severity of the patient's condition at admission to the hospital. Included in this category are comorbidities, additional conditions that may exacerbate the condition that is the major reason for admission. The LOS is expected to be longer for more severely ill patients and for patients with comorbidities.

B. Demographic Variables describe the patient's age, sex, and race. The effect of demographic factors on LOS for patients with pediatric asthma is not known.

C. Hospital Variables describe the characteristics of the hospital in which the patient was treated. It is expected that LOS will be lower for hospitals with more services and with a higher staff-to-bed ratio. The effects of size of hospital, measured by the number of beds in the hospital (bedsize), and the categorization of that hospital as a teaching hospital or not (teachcat) have not been explored. In this case study these variables are named bedsize and teachcat.

D. Treatment Variables describe aspects of the patient's treatment during hospitalization. Treatment variables are not fully known until the hospital stay is complete. It is expected that LOS will be longer for patients who have treatment complications and process of care failures (the hospital fails to adhere to standards of care given the patient's condition). Finally, patients with more diagnostic tests probably are more severely ill. Consequently, LOS is expected to increase with the number of tests performed.

QUESTIONS OF INTEREST

•   Are there differences in LOS between the insurers?
•   Do the differences hold up once you have accounted for differences in hospital and patient characteristics?
•   What are the differences in patient severity characteristics, demographics, hospital characteristics, and treatment variables between the two insurers?

DATA

The data in this case study were analyzed using the Statistical Analysis System (SAS). Below see the SAS Input Statement for the Pediatric Asthma Data found in the file Case05.txt.

```
INPUT    LOS  1-2  HOSPITAL  4-6  PATIENT  8-12  INSURER  14-15
SUMCOMRB  17-18  CMPL11  21    CMPL12  23    SEVMOD  25  SEVSEV  27
HISTF01  29-30   HISTF02  32  HISTF03  34-35  AGE  37-38  FEMALE  40
RACE  42    SUMSERV  44-46  BEDSIZE  48    OWNER  50  TEACHCAT  52
ANYCOMP  54  HIGHPOC  56  MEDPOC  58    LOWPOC  60    DIAGTSTS  62-63
@65  FTETOBED  7.4   @73  PCTINS1  7.4;
```

Here are the first four observations for hospital 17.

```
        H    S                           T              D        F
        O   P I U           H H H        S B    E A H        I   T      P
        S   A N M C C S S I I I     F    U E    A N I M L A  E   C
        P   T S C M M E E S S S     E    M D O C Y G E O G   T   T
        I   I U O P P V V T T T   M R    S S W H C H D W T   O   I
   O L  T   E R M L L M S F F F   A A A  E I N C O P P P S   B   N
   B O  A   N E R 1 1 O E 0 0 0   G L C  R Z E A M O O O T   E   S
   S S  L   T R B 1 2 D V 1 2 3   E E E  V E R T P C C C S   D   1

   1 1 17  10 0 0 0 0 1 0 0 0 1  10 1 2  51 4 2 1 0 1 0 0 1  4.64 11.52
   2 2 17  11 0 0 0 0 1 0 0 0 0   6 0 1  51 4 2 1 0 1 0 0 2  4.64 11.52
   3 3 17  12 0 0 0 1 1 0 0 0 0   3 0 1  51 4 2 1 0 0 2 0 3  4.64 11.52
   4 4 17  13 0 0 0 0 1 1 0 0 0  11 0 3  51 4 2 1 0 0 0 1 3  4.64 11.52
```

Table 1 lists the variables along with their ranges of values.

ANALYSIS

A major objective of the analysis is testing whether the average LOS is different between the two insurers. The unadjusted average LOS is higher by 0.6 days for Insurer A than it is for Insurer B ($p < .001$). To test whether this difference in average LOS can be attributed to differences in patient and hospital characteristics or between the two insurers, a regression model will be applied, with LOS as the dependent variable.

The approach will be to fit the model in four stages. In each stage there will be at least two steps. In the first step all variables from one of the categories A through D as shown below will be added and their contribution to the model assessed. In the second step a subset of the variables selected from the first step will be analyzed. In Tables 2 and 3 check the means, standard deviations, and frequencies for each variable and compare Insurers A and B. For which of the variables are there statistically significant differences given at the 0.05 level of significance? Are there any surprising differences or lack of differences shown in the means and contingency tables? Construct boxplots comparing LOS and DIAGTSTS for the two insurers. What do the boxplots reveal about the shape of the distributions for these two variables? What effect might the characteristics observed have on a regression analysis of LOS?

Next check the correlations among variables in Table 4. Are there any unusually high correlations among variables? Are there significantly high intercorrelations among the independent variables? Recall that LOS is the dependent variable.

Recall that the variables fall into four major categories:

Category	Variable Description
A.	Patient Severity Variables
B.	Demographic Variables
C.	Hospital Level Variables
D.	Treatment Variables

Begin building the regression model. Since the objective is to compare insurers, Insurer will be included in every model. At each stage check the R^2, the overall significance level, the significance for each of the included variables, the value of the regression coefficient, and the variance inflation factors (VIFs). Eliminate variables that are not significant at the 0.10 level. For variables that have high VIFs consider whether you might eliminate one or more to lower the multicollinearity [Mendenhall and Sincich, 1993]. Rerun the model with the selected variables and check the R^2 again.

Table 1. *Definition of variables.*

VARIABLE NAME	DESCRIPTION	VALUES
INSURER	Insurer who Pays the Hospital Bill	0 = A, 1 = B
Patient Severity		
SEVMOD	Asthma Severity is Moderate or Higher	1 = Yes, 0 = No
SEVSEV	Asthma Severity is Severe	1 = Yes, 0 = No
CMPL11	Bronchitis Present?	1 = Yes, 0 = No
CMPL12	Pneumonia Present?	1 = Yes, 0 = No
SUMCOMRB	Number of Comorbidities	Nonnegative Integers
HISTF01	History of Respiratory Failure	1 = Yes, 0 = No
HISTF02	Oral Steroid Dependent	1 = Yes, 0 = No
HISTF03	Two or More Previous Hospitalizations for Asthma	1 = Yes, 0 = No
Patient Demographics		
AGE	Patient Age in Years	Integers 2–18
FEMALE	Patient is Female	1 = Yes, 0 = No
RACE	Patient Race	1 = White
		2 = Hispanic
		3 = Black
		4 = Asian/Pacific Islander,
		5 = Unknown
Hospital Characteristics		
OWNER	Ownership Category	1 = Public,
		2 = Private
NUMSERV	Number of Hospital Services	Positive Integers
BEDSIZE	Hospital Bedsize Category	1 = 1–99,
		2 = 100–249,
		3 = 250–400,
		4 = 401–650
FTETOBED	Full Time Staff per Bed	Positive Real Numbers
PCTINS1	Percent of Annual Patients Insured by Insurer B	Real Numbers 0–100
TEACHCAT	Degree of Teaching Category	0 = None,
		1 = Minor,
		2 = Major
Treatment Variables		
ANYCOMP	Any Treatment Complications	1 = Yes, 0 = No
DIAGTSTS	Number of Diagnostic Tests Ordered or Performed	Nonnegative Integers
LOWPOC	Number of Low Level Process of Care Failures	Nonnegative Integers
MEDPOC	Number of Medium Level Process of Care Failures	Nonnegative Integers
HIGHPOC	Number of High Level Process of Care Failures	Nonnegative Integers

Table 2. *Descriptive statistics: mean and standard error.*

Variables	Insurer A	Insurer B	*p*-value
Length of Stay in Days	2.3	2.9	<.001
(mean)LOS	(.06)	(.08)	
Severity > 2 (%)	78.4	82.3	.163
SEVMOD	(2.1)	(1.9)	
Severity > 3 (%)	12.5	8.6	.076
SEVSEV	(1.7)	(1.4)	
Number of Comorbidities	0.06	0.14	0.001
SUMCOMRB	(.01)	(.02)	
Bronchitis (%)	6.1	8.6	0.183
CMPL11	(1.2)	(1.4)	
Pneumonia	17.3	15.9	0.599
CMPL12	(1.9)	(1.8)	
History Respiratory	2.3	3.0	0.519
Failure (%) HISTF01	(0.8)	(0.8)	
Oral Steroid HISTF02	1.0	1.5	0.533
Dependent (%)	(0.5)	(0.6)	
>2 Previous HISTF03	24.4	37.6	<.001
Hospitalization (%)	(2.2)	(2.4)	
Age in years (mean)	6.9	6.6	.341
AGE	(0.2)	(0.2)	
Female (%)	37.4	34.6	.412
FEMALE	(2.4)	(2.4)	
Number of Services	36.4	35.9	0.500
SUMSERV	(0.5)	(0.6)	
Private Hospital (%)	93.6	73.0	<.001
PRIVATE	(0.2)	(0.4)	
Ratio of FTEs to Beds	4.6	6.1	<.001
FTETOBED	(0.9)	(1.1)	
Percent Payer 1	15.2	39.7	<.001
PCTINS1	(18.4)	(21.6)	
Any Treatment ANYCOMP	2.0	2.0	0.988
Complication (%)	(0.7)	(0.7)	
# High Risk Care Problems	0.28	0.31	0.547
(mean) HIGHPOC)	(.03)	(.03)	
# Medium Risk Care MEDPOC	0.61	0.57	0.436
Problems (mean)	(0.4)	(0.4)	
# Low Risk Care Problems	0.68	0.81	0.042
LOWPOC	(.04)	(.04)	
# Diagnostic Tests	3.19	3.07	0.151
(mean) DIAGTSTS	(.05)	(.06)	

Table 3. *Contingency tables.*

Patient/Hospital Variables	Insurer A	Insurer B
Race (%) *p*<.001		
Black	28.5	41.2
White	37.2	18.2
Hispanic	19.1	33.3
Asian/Pacific Islander	8.1	4.8
Unknown	7.1	2.5
Hospital Beds (%) *p*<.001		
1–99	2.8	0.8
100–249	38.7	50.8
250–400	46.1	32.1
401–650	12.5	16.4
Teaching Status (%) *p*<.001		
Nonteaching	55.5	50.0
Minor Teaching	33.8	15.7
Major Teaching	10.7	34.3

Based on Chi-square tests of association

Table 4A. *Correlation analysis.*

Pearson Correlation Coefficients / Prob > |*R*| under *Ho*: Rho = 0 / *N* = 789

LOS With

DIAGTSTS	INSURER	TEACHING	CMPL	MINORTCH	CMPL12
0.32813	0.20392	-0.19663	0.17702	-0.17129	0.14708
0.0001	0.0001	0.0001	0.0001	0.0001	0.0001

SEVMOD	HISTF03	SEVSEV	BEDGT400	HISTF01	AGE
0.13591	0.11407	0.11130	-0.11048	0.10950	0.10450
0.0001	0.0013	0.0017	0.0019	0.0021	0.0033

CMPL11	SUMSERV	ANYCOMP	BEDGT249	FTETOBED	MAJORTCH
0.09180	-0.09172	0.08815	-0.06857	-0.06803	-0.05809
0.0099	0.0099	0.0133	0.0542	0.0561	0.1030

SUMCOMR B	LOWPOC	MEDPOC	PCTINS1	HIGHPOC	HISTF02
0.05427	0.04752	-0.04349	0.03735	-0.02797	-0.00911
0.1278	0.1824	0.2224	0.2947	0.4328	0.7984

FEMALE	PRIVATE
-0.00720	-0.00398
0.8399	0.9111

Table 4B. *Correlation analysis.*

Pearson Correlation Coefficients / Prob > |R| under *Ho*: Rho = 0 / *N* = 789

	LOS	INSURER	SUMCOMRB	CMPL11	CMPL12
LOS	1.0000.0	0.20392	0.05427	0.09180	0.14708
	0.0	0.0001	0.1278	0.0099	0.0001
INSURER	0.20392	1.00000	0.11545	0.04749	-0.01873
	0.0001	0.0	0.0012	0.1826	0.5994
SUMCOMRB	0.05427	0.11545	1.00000	0.11130	-0.02560
	0.1278	0.0012	0.0	0.0017	0.4727
CMPL11	0.09180	0.04749	0.11130	1.00000	-0.03432
	0.0099	0.1826	0.0017	0.0	0.3356
CMPL12	0.14708	-0.01873	-0.02560	-0.03432	1.00000
	0.0001	0.5994	0.4727	0.3356	0.0
CMPL	0.17702	0.02199	0.05263	0.51442	0.81486
	0.0001	0.5374	0.1397	0.0001	0.0001
HISTF01	0.10950	0.02299	0.10905	0.01377	0.01086
	0.0021	0.5190	0.0022	0.6994	0.7606
HISTF02	-0.00911	0.02223	-0.00128	-0.03191	-0.05055
	0.7984	0.5330	0.9714	0.3707	0.1560
HISTF03	0.11407	0.14262	0.08297	0.01039	-0.03444
	0.0013	0.0001	0.0198	0.7708	0.3340

	CMPL	HISTF01	HISTF02	HISTF03
LOS	0.17702	0.10950	-0.00911	0.11407
	0.0001	0.0021	0.7984	0.0013
INSURER	0.02199	0.02299	0.02223	0.14262
	0.5374	0.5190	0.5330	0.0001
SUMCOMRB	0.05263	0.10905	-0.00128	0.08297
	0.1397	0.0022	0.9714	0.0198
CMPL11	0.51442	0.01377	-0.03191	0.01039
	0.0001	0.6994	0.3707	0.7708
CMPL12	0.81486	0.01086	-0.05055	-0.03444
	0.0001	0.7606	0.1560	0.3340
CMPL	1.00000	0.02161	-0.06204	-0.02285
	0.0	0.5445	0.0816	0.5215
HISTF01	0.02161	1.00000	-0.01874	0.19535
	0.5445	0.0	0.5993	0.0001
HISTF02	-0.06204	-0.01874	1.00000	0.09537
	0.0816	0.5993	0.0	0.0073
HISTF03	-0.02285	0.19535	0.09537	1.00000
	0.5215	0.0001	0.0073	0.0

Table 4C. *Correlation analysis.*

Pearson Correlation Coefficients / Prob > |*R*| under *Ho*: Rho = 0 / *N* = 789

	LOS	ANYCOMP	PCTINS1	PRIVATE	MINORTCH
LOS	1.00000	0.08815	0.03735	−0.00398	−0.17129
	0.0	0.0133	0.2947	0.9111	0.0001
ANYCOMP	0.08815	1.00000	0.01376	−0.00779	0.02180
	0.0133	0.0	0.6995	0.8271	0.5410
PCTINS1	0.03735	0.01376	1.00000	−0.38699	−0.39785
	0.2947	0.6995	0.0	0.0001	0.0001
PRIVATE	−0.00398	−0.00779	−0.38699	1.00000	0.07576
	0.9111	0.8271	0.0001	0.0	0.0334
MINORTCH	−0.17129	0.02180	−0.39785	0.07576	1.00000
	0.0001	0.5410	0.0001	0.0334	0.0001
TEACHING	−0.19663	0.04387	0.11624	−0.25572	0.60508
	0.0001	0.2183	0.0011	0.0001	0.0001
MAJORTCH	−0.05809	0.02991	0.54942	−0.38363	−0.30925
	0.1030	0.4015	0.0001	0.0001	0.0001
BEDGT249	−0.06857	0.02600	−0.22419	−0.09803	0.23389
	0.0542	0.4658	0.0001	0.0059	0.0001
BEDGT400	−0.11048	0.01760	0.05648	−0.32769	0.02361
	0.0019	0.6215	0.1129	0.0001	0.5078

	TEACHING	MAJORTCH	BEDGT249	BEDGT400
LOS	0.19663	−0.05809	−0.06857	−0.11048
	0.0001	0.1030	0.0542	0.0019
ANYCOMP	0.04387	0.02991	0.02600	0.01760
	0.2183	0.4015	0.4658	0.6215
PCTINS1	0.11624	0.54942	−0.22419	0.05648
	0.0011	0.0001	0.0001	0.1129
PRIVATE	−0.25572	−0.38363	−0.09803	−0.32769
	0.0001	0.0001	0.0059	0.0001
MINORTCH	0.60508	−0.30925	0.23389	0.02361
	0.0001	0.0001	0.0001	0.5078
TEACHING	1.00000	0.57001	0.32828	0.43400
	0.0	0.0001	0.0001	0.0001
MAJORTCH	0.57001	1.00000	0.15074	0.49403
	0.0001	0.0	0.0001	0.0001
BEDGT249	0.32828	0.15074	1.00000	0.38325
	0.0001	0.0001	0.0	0.0001
BEDGT400	0.43400	0.49403	0.38325	1.00000
	0.0001	0.0001	0.0001	0.0

Do the regression analysis for the category A variables. What variables did you retain for the model?

Now, add the Demographic (or category B) variables to the ones selected from the Patient Severity variables. As each stage is completed, do not drop a variable selected at that stage from the set at a later stage; i.e., build an additive model. Again, check your observed significance levels and VIFs. Run the model with the selected variables. Do any of the category A variables become insignificant at the 0.10 level? If so, should they be dropped from the model? Why?

As the next step, add the Hospital Variables. Answer the same questions as for the first two categories.

Finally, add the Treatment Variables. Plot the residuals against DIAGTSTS. Is there a pattern?

At the final stage add the term for the square of DIAGTSTS. Squaring terms can introduce problems with multicollinearity; such multicollinearity caused by squaring terms is called structural. In cases of structural multicollinearity more accurate estimates of the regression coefficients can often be found if the variable is first standardized or centered [Mendenhall and Sincich, 1993]. Is there evidence of structural multicollinearity here? Center the variable DIAGTSTS (i.e., subtract the mean and divide by the standard deviation) and square the centered variable. Rerun the model with the centered variables. Is there still evidence of structural multicollinearity? Compare the residual plots for the centered and uncentered models. What effect did centering the variable have on the regression coefficients and R^2?

Compare the R^2 values for the four stages. Check the regression diagnostics, including plots of the residuals; check the residuals for normality and for influential observations and outliers. If there are outliers identified by the studentized residuals, the hat diagonal, the dffits or the dfbetas [Neter, Wasserman, and Kutner, 1990], try to identify the reason for the unusual value. Try running the model excluding the outliers or influential observations and check the effects on the coefficients, the residuals, and the R^2.

This four-stage approach defines a framework in which severity takes precedence over demographics, demographics take precedence over hospital variables, and hospital variables take precedence over treatment variables in explaining LOS variability. The final model at each stage adjusts the effect of Insurer for variables that are increasingly easier to control. The patient's severity of asthma at admission is most often outside anybody's control. Patient demographics are fixed, although they can be controlled in the aggregate to a certain extent through insurance enrollment policies. The insurer can control hospital characteristics by selectively contracting with hospital providers and then by forcing the insured to receive nonemergency treatment only at contracting hospitals. The hospital can lower the rates of treatment complications and process of care failures by increasing staff quality and by adopting strict practice standards. Standard treatment protocols demand that the hospital perform certain diagnostic tests. Consequently, although the hospital has control over the number of diagnostic tests performed for a patient, this measure is most likely a proxy for overall patient severity of illness, rather than something the hospital can use to reduce LOS.

What conclusion about the difference between insurers do you draw on the basis of the analysis? Once the model has accounted for differences in the patients such as severity of condition, previous hospitalizations, and comorbidities, are there still differences between insurers? Has the addition of other variables changed the interpretation of the difference between insurers?

Try rerunning the analyses by looking at each category of variable separately, but always including Insurer. Which of the categories provides the greatest amount of explanatory value? What set of variables would you use if you chose the variables selected from each category independently? Is this model noticeably different from the additive or phased approach you tried first?

Compare the method you were using for variable selection to software package programs that do stepwise regression. Try running the same variables using forward, backward, and stepwise methods. To gain comparability with your approach in the phased (or additive) method use a significance level 0.10.

REFERENCES

Glantz, Stanton A. and Bryan K. Slinker. *Primer of Applied Regression and Analysis of Variance*. New York: McGraw–Hill, Inc., 1990.

Mendenhall, William, and Terry Sincich. *A Second Course in Business Statistics*. New York: Dellen, 1993.

Neter, John, William Wasserman, and Michael H. Kutner. *Applied Linear Statistical Models*, Third Edition. Burr Ridge, Illinois: Irwin, 1990.

BIOGRAPHIES

Nancy Schoeps teaches statistics at the University of North Carolina at Charlotte and directs the university's Office of Statistics and Applied Mathematics (OSAM). OSAM provides statistical consulting services for the university and for the business and industrial community. She received her doctorate from Syracuse University.

Robert Houchens received his Ph.D. in Applied Statistics from the University of California, Riverside. During his graduate studies, Dr. Houchens worked as a statistician for the RAND Corporation on projects in the areas of military strategy, manpower planning, criminal justice, and civil justice. For the last 14 years Dr. Houchens has been employed as a statistician for The MEDSTAT Group, planning and conducting studies in health services research.

CHAPTER 6

COMPARING NONSTERIODAL ANTI-INFLAMMATORY DRUGS WITH RESPECT TO STOMACH DAMAGE

Tom Filloon and Jack Tubbs

This case study may seem to be a rather trivial exercise, but we feel that it contains many of the important ideas that applied statisticians use on a day-to-day basis. It discusses the quantile–quantile (QQ) plot, normality assumptions, comparing distributions, and calculating p-values. Furthermore, it shows the great utility of the Mann–Whitney–Wilcoxon rank sum approach.

INTRODUCTION

Many people take medication daily for the treatment of arthritis. Painful, swollen joints are a source of problems for arthritis sufferers. Pain relief and anti-inflammatory benefits can be achieved by drugs classified as NSAIDs (NonSteroidal Anti-Inflammatory Drugs), which include such drugs as ibuprofen (Motrin). One potential side effect with the long-term use of this class of drugs is that they can possibly cause severe stomach damage (lesions, ulcers, perforation, death). In addition, if a person has developed a stomach ulcer, then this type of drug has the potential for delaying the time it takes for an ulcer to heal. The goal of a pharmaceutical company's research is to provide a better, safer drug for treating arthritis (i.e., developing an arthritis drug that does not slow the ulcer healing process). In this study, we are evaluating two drugs in an animal ulcer healing experiment in an effort to determine a new, more stomach-safe NSAID for use by arthritis sufferers. Analysis of this data will include descriptive statistics, assessing normality, permutation testing, and sample size determination.

BACKGROUND INFORMATION

An animal (rat) experimental model has been developed to evaluate NSAIDs with regard to their effects on ulcer healing. In this animal model, all animals are given a large dose of a known stomach-damaging compound. It has been shown that after approximately 2 weeks, the majority of the stomach damage created by the initial insult is gone (i.e., substantial ulcer healing has taken place). In an effort to mimic how people take pain medication for arthritis, an experimental compound is administered to each rat twice daily for these 14 days. At the end of this treatment period, the animals are sacrificed and their stomachs are examined in order to quantify the amount of damage present as indicated by long, narrow indentations called lesions. The response variable of interest is the sum of the lesion lengths in millimeters (mm).

This type of experiment has been run several times previously in order to compare the amount of stomach damage remaining for the standard drug-treated animals (denoted as OLD) as compared to the control-treated animals (PLACEBO). That is, one has observed and would expect to observe larger lesion lengths in the Old group as compared with the Placebo group at the end of 14 days. Hence, one can conclude that the Old treatment is delaying ulcer healing as there is more damage left at the day 14 evaluation.

In the current study described here, a new drug treatment (NEW) will also be used with the Old and Placebo treatments. The purpose of this study is investigate how the two NSAIDs (Old, New) compare with the Placebo group in order to determine if these drugs delay the ulcer healing process. Additionally, it is of interest to compare the NSAIDs to one another.

QUESTIONS OF INTEREST

There are several questions of interest that this study will address.
1. Does the Old treatment delay ulcer healing as compared to the Placebo? This would indicate that the results of this study are similar to previous experiments.
2. Does the New treatment also delay healing as compared with the Placebo?
3. Is the New Treatment superior to the Old treatment?

DATA

From prior data and power calculations, it was determined that 25–30 animals were needed for each treatment in order to have sufficient power to be able to detect differences between treatments. To account for expected accidental loss of animals through the course of the study, 35 animals were randomized to each of the 3 groups. As this number of animals was too large to perform the experiment's final day procedures in one day, a complete block design was used. Four blocks of animals were used with each block of animals starting the study on a different day such that they were staggered over days in order to balance out any potential time trends. As is usual with animal experiments, a small proportion of the animals died during the course of the study, causing final sample sizes to range from 32–34 in the 3 groups. Table 1 shows a partial data listing. The complete data set for the 3 treatment groups is found in file Case06.txt.

SUGGESTED ANALYSIS

1. Use simple descriptive procedures for each of the treatment groups in order to visualize the data and to determine how the three groups' means and medians compare. Based upon these simple procedures, what preliminary results would you expect? What problem areas do you observe with the data?
   a. Determine if the data are normally distributed. If not, what procedures would you use to achieve normality? How do these procedures work in these data?
   b. Although there are three treatments in the study the main interest is found in the pairwise comparisons of the groups. Test to determine the nature of the differences using two sample t-tests. Are you justified in using this approach? If so, why? If not, why not?

c.   An alternative test for comparing two treatments is called the Wilcoxon rank sum test. Perform a Wilcoxon rank sum analysis for the three pairwise treatment comparisons. What are your conclusions? How do these conclusions compare with your results found in #3? How do you know whether or not either set of conclusions are correct for these data?

d.   A method called permutation (or randomization) tests can be used to obtain p-values for any test of hypothesis. What are the results found in using a permutation test? What are your overall conclusions concerning the three treatment groups' means or medians?

e.   The statistic used in the Wilcoxon ranked sum test can be used to estimate Prob($Y > X$). Find the estimates of these probabilities for the three pairwise comparisons.

f.   Results from previous studies were used to determine sample sizes for this study. Use the information from this study (obtain effect size from observed Old vs. Placebo comparison) to determine the sample sizes for similar analysis in a subsequent experiment.

Table 1. *Listing of data.*

Block/Day	Treatment	Lesion Length (mm)
1	Placebo	0.00
1	New	3.16
1	Old	0.00
1	Old	8.00
1	Old	8.21
1	New	0.04
1	Old	9.90
1	Placebo	0.36
1	New	8.23
1	Placebo	0.77
1	New	0.00
.	.	.
.	.	.
.	.	.
4	Old	5.97
4	New	0.90
4	New	2.86
4	New	1.64

INSTRUCTIONS FOR PRESENTATION OF RESULTS

Results should be summarized in a written report with graphs. Optionally, oral presentations could be given from the written reports.

REFERENCES

Boos, D.D. and Brownie, C. (1992), "A Rank Based Mixed Model Approach to Multisite Clinical Trials," Biometrics 48, 61–72.

Hollander, M. and Wolfe, D.A. (1973), Nonparametric Statistical Methods, John Wiley & Sons, New York.

Lehr, R. (1992), "Sixteen S-Squared over D-Squared: A Relation for Crude Sample Size Estimates," Statistics in Medicine 11, 1099–1102.

Monti, K.L. (1995), "Folded Empirical Distribution Functions - Mountain Plots," The American Statistician 49, 342–345.

Noether, G.E. (1987), "Sample Size Determination for Some Common Nonparametric Tests," Journal of the American Statistical Association 82, 645–647.

BIOGRAPHIES

Tom Filloon is a Senior Statistician in the Biostatistics & Medical Surveillance Department at The Procter and Gamble Company. He has over ten years experience working in the biopharmaceutical and health care areas. Over the past several years, he has used robust and nonparametric data analysis extensively in both preclinical and clinical research programs.

Jack Tubbs is a Professor in the Department of Mathematical Sciences at the Univeristy of Arkansas. He has over twenty years of teaching experience at both the undergraduate and graduate level. His special interests are in classification and computer vision. Lately, he has had an interest in using real-world problems to motivate statistical methodology and problem solving.

CHAPTER 7

VALIDATING AN ASSAY OF VIRAL CONTAMINATION

Lawrence I-Kuei Lin and W. Robert Stephenson

Viral contamination is of great concern to the makers, and users, of biological products such as blood clotting Factor Eight (given to people with hemophilia) and human blood substitute (a product still in development). How does one guarantee that such products are free of viral contamination? The first step is to have an assay that can accurately and precisely measure viral contamination. An assay is an analysis of a substance to determine the presence or absence of a specific ingredient. Most of you will be familiar with the idea of an assay of mineral ore to determine the amount of gold. In a viral assay, a solution is analyzed to determine the presence or absence of a specific virus. A viral assay can also be used to determine the amount of virus in the solution, the total viral titer. In order to ensure the accuracy and precision of an assay, it must be validated. The validation of an assay has three components: linearity (or proportionality), precision, and sensitivity. Each of these components requires the use of statistical methods. This case study looks at validating a viral assay using bovine viral diarrhea virus (BVDV). Other methods are used to validate viral assays for human immunodeficiency virus (HIV), the virus that causes AIDS.

INTRODUCTION

In order to validate an assay one must start with something that has a known viral contamination. To do this, virologists spike a sterile stock solution with a known amount of a particular virus, in our case BVDV. BVDV is a virus that affects the gastrointestinal system of cattle causing severe diarrhea. The virus is particularly harmful to pregnant cattle because of its ability to infect the fetus. BVDV is closely related to the hog cholera virus and a similar virus that affects sheep. The BVDV has the property that when cultured in a petri dish the viral particles form plaques, circular regions in the culture medium. These plaques are easily visible under a microscope or to the naked eye when a stain is used. Each plaque is associated with a single viral particle. By counting the number of distinct plaques, so-called plaque forming units (PFUs), per mL of volume, one can estimate the total viral titer. HIV does not form plaques so a different technique that looks for simply the presence or absence of the virus is used.

BACKGROUND INFORMATION

For BVDV it seems straightforward to determine the total viral titer. However, with a large enough concentration of virus, the entire petri dish becomes one large plaque, and it is impossible to count individual PFUs. In order to produce countable plaques one must dilute the spiked solution. The dilution is performed as follows. One mL of the spiked solution is mixed with 99,999 mL of sterile solution, a reduction to 1 in 100,000. This dilute-contaminated solution is further diluted in a series of steps, a serial dilution. At the first dilution 1 part of the dilute-contaminated solution is mixed with 2 parts of sterile solution. At the second dilution 1 part of the first dilution is mixed with 2 parts of the sterile solution. This continues in the same manner so that at dilution d, 1 part from dilution $d-1$ is mixed with 2 parts of sterile solution. At each step in the serial dilution the contamination should be 1/3 as great as in the previous step. At each step, 4 petri dishes are prepared with material from that dilution. For the first several dilutions, the petri dishes produce uncountable plaques because of the overabundance of viral particles. In order to proceed, we need at least 2 dilutions that yield countable (but nonzero) plaques.

The process then is to spike, dilute, culture, and count. Since variability can affect each of the steps in the process, the process is repeated several times (called samples).

QUESTIONS OF INTEREST

In general, one wishes to know if the method of assay is valid; i.e., is it sufficiently accurate and precise to use in routine laboratory work? The validation of an assay method has three components: linearity (or proportionality), precision, and sensitivity. This case study looks at only the first two components.

Linearity/Proportionality

Given the counts of the PFUs/mL from the serial dilutions, can we estimate the total viral titer in the undiluted contaminated stock?

Precision

As with all data there is variation in response, in this case variation in the number of PFUs/mL. What causes the majority of this variation? Is it attributable to differences from sample to sample and/or to differences within a sample? How does this affect our estimate of the total viral titer?

DATA: PFU ASSAY VALIDATION: BVDV

Name of Data File: Case07.txt
There are nine samples, resulting from nine runs of the process of spike, dilute, culture, and count. Each sample has seven dilutions and four petri dish counts of PFUs/mL. Portions of the full data set necessary for the basic analysis are given in Tables 1 and 2.

Linearity/Proportionality

Table 1. *Number of PFUs/mL for sample* 1, *dilutions* 3 *through* 9.

Sample	Dilution	PFUs/mL			
1	3	75	66	84	82
1	4	24	18	27	21
1	5	4	6	8	7
1	6	4	1	2	2
1	7	0	0	0	1
1	8	0	2	0	1
1	9	0	0	0	0

Note: Dilutions below 3 yield uncountable PFUs/mL due to the high concentration of virus.
Note: The full data set has a total of nine samples.

Precision

Table 2. *Number of PFUs/mL for dilutions* 3, *samples* 1 *through* 9.

Sample	Dilution	PFUs/mL			
1	3	75	66	84	82
2	3	55	43	33	35
3	3	63	57	54	65
4	3	105	70	87	73
5	3	63	59	66	61
6	3	39	47	46	53
7	3	36	38	52	38
8	3	45	56	34	30
9	3	30	46	45	49

Note: The full data set includes dilutions 3, 4, 5, 6, 7, 8, and 9.

ANALYSIS

Linearity/Proportionality

For sample 1 plot the number of PFUs/mL versus the dilution. Describe the general relationship between dilution and the number of PFUs/mL. Will a straight line give a good approximation to the relationship between dilution and the number of PFUs/mL?

Recall that at each dilution there should be approximately 1/3 as much viral contamination as at the previous dilution. That is, the number of PFUs/mL should be falling in a multiplicative fashion. Taking the logarithm of the number of PFUs/mL can change this multiplicative relationship into an additive one. The logarithm, base 10, of a number is simply the power of 10 that will give you that number. For example, the logarithm, base 10, of 100 is 2 since $10^2 = 100$. The logarithms of the PFUs/mL for sample 1 appear in Table 3.

Table 3. *Logarithm of PFUs/mL for sample 1, dilutions 3 through 9.*

Sample	Dilution	Y=log$_{10}$(PFUs/mL)			
1	3	1.875	1.820	1.924	1.914
1	4	1.380	1.255	1.431	1.322
1	5	0.602	0.778	0.903	0.845
1	6	0.602	0.000	0.301	0.301
1	7	*	*	*	0.000
1	8	*	0.301	*	0.000
1	9	*	*	*	*

Note: If PFUs/mL = 0, then log$_{10}$(PFUs/mL) is undefined and is denoted by the asterisk.

Again for sample 1, plot the response $Y=\log_{10}$(PFUs/mL) vs. dilution. On your plot note the zero PFUs/mL counts, the asterisks, in some way. Describe the general relationship between dilution and this response. Does the relationship appear to be linear for dilutions 3, 4, 5, and 6? Compute the least squares line relating $Y=\log_{10}$(PFUs/mL) to dilution using only dilutions 3, 4, 5, and 6. Give the value of the coefficient of determination, R^2, and comment on what this value tells you about the fit of the straight line. Give an interpretation of each of the estimated coefficients (the intercept and the slope) within the context of the problem. An estimate of the total viral titer, the number of viral particles in the undiluted contaminated solution, is

$$10^{(\hat{Y}_{d=0}-1)} \text{ million particles per mL,}$$

where $\hat{Y}_{d=0}$ is the predicted value of Y when the dilution is zero. By using $\hat{Y}_{d=0}-1$ in the exponent, the titer is given in millions of particles rather than hundreds of thousands of particles. Using the regression equation you developed, estimate the total viral titer. Calculate the residuals and plot these against the dilution. Describe the pattern in this residual plot.

This analysis can be repeated for the other samples in the data set.

Precision

For each dilution there is variability within the four observations. There is also variation from sample to sample. Which of these sources of variation contributes the most to the overall variation? For dilution 3 consider the one-way classification according to sample. Plot the data to show the variation within samples and among samples. Analyze these data with an ANOVA to separate the within (intra) sample variation from the among (inter) sample variation. Table 4 gives the expected mean squares which may be helpful in estimating the variance components.

Table 4. *Expected mean squares and variance components.*

Source	Expected Mean Square	Variance Component	
Samples	$\sigma^2 + 4\sigma^2_{samples}$	Intersample:	$\sigma^2_{samples}$
Error	σ^2	Intrasample:	σ^2
Total:		Total:	$\sigma^2 + \sigma^2_{samples}$

What percentage of the total variance is attributable to intersample variation? What percentage of the total variance is attributable to intrasample variation? If Intrasample variation is too large, larger than the intersample variation, then the precision of the data is suspect. In that case the data for that dilution should not be used to predict the total viral concentration. Another way to quantify precision, or lack thereof, is through the coefficient of variation, CV. The coefficient of variation is defined to be the square root of the total variance divided by the mean. It is often expressed as a percentage. As a rule of thumb, if the CV% is greater than 50% there is too much variability to accurately predict the total viral concentration.

Compute the CV% for dilution 3. Repeat the analysis with dilutions 4, 5, 6, and 7. For which dilution(s) is the CV% greater than 50%? How does this affect your assessment of linearity?

INSTRUCTIONS FOR PRESENTATION OF RESULTS

Most projects require presentation of results to clients. Presentations have more impact than documents alone. To be successful a presentation should be

- In terms of the client's needs
- In tune with the client's values
- In the client's language

The statistician is responsible for ensuring this. She must translate her thinking into words understandable to the client. Some possible client audiences for this case might be

- Manager of statisticians and staff
- Manager of manufacturing and staff manager of virologists and staff
- Chief executive of organization and staff

Linearity/Proportionality

Prepare a presentation of your findings. Direct your presentation to one of the possible client audiences described above. Be sure to include graphs of the data. An important consideration is the final estimate of the total virus titer; however, this is only as good as the regression equation used to estimate it. Comment on the adequacy of the linear regression that relates the log transformed PFUs/mL to the dilution.

Precision

Prepare a presentation of your findings directed at one of the possible client audiences described above. This presentation should indicate which dilutions provide reliable data for estimating total virus titer. Include any summary tables and graphs you feel are appropriate.

REFERENCES

Lin, Lawrence I-Kuei (1992), "Assay Validation Using the Concordance Correlation Coefficient," Biometrics, 48, pp. 599–604.

Chow, Shein-Chung and Tse, Siu-Keung (1991), "On the Estimation of Total Variability in Assay Validation," Statistics in Medicine,10, pp.1543–1553.

BIOGRAPHIES

Lawrence I-Kuei Lin is an applied statistician with 20 years of consulting experience, mostly with Baxter Healthcare Co. He has worked on projects in clinical trials, product development, reliability, laboratory quality control monitoring, assay validation, and customer requirements. He has developed and patented an interlaboratory quality control super highway system. In addition to his work at Baxter Healthcare he is an Adjunct Professor at the University of Illinois at Chicago. Dr. Lin served as the President of the Northern Illinois Chapter of the American Statistical Association (NIC-ASA) in 1994.

W. Robert Stephenson is a Professor of Statistics at Iowa State University. At Iowa State he is heavily involved in teaching the introductory and intermediate level applied statistics courses. In addition, he has developed a two-semester sequence of courses, Applied Statistics for Industry, for the General Motors Technical Education Program. These courses are taught on campus and delivered to General Motors sites in Michigan, Ohio, Arizona, and Mexico, as well as industrial sites in Iowa. Professor Stephenson serves as an Associate Editor of STATS: The Magazine for Students of Statistics and Technometrics. He is a Fellow of the American Statistical Association.

CHAPTER 8

CONTROL CHARTS FOR QUALITY CHARACTERISTICS UNDER NONNORMAL DISTRIBUTIONS

Youn-Min Chou, Galen D. Halverson, and Steve T. Mandraccia

When using Shewhart control charts, the underlying distribution of the quality characteristic must be at least approximately normal. In many processes, the assumption of normality is violated or unjustifiable. If the quality characteristic is not so distributed, then the control limits may be entirely inappropriate, and we may be seriously misled by using these control charts. In this case study, we discuss several "state of the art" curve-fitting methods for improving the technical validity of control charts for the nonnormal situation. We also compare their practical application qualities using data from the semiconductor industry.

INTRODUCTION

To set up a control chart for the particle counts, the control limits may be calculated according to historical data from the particle count database. Two frequently used charts for particle counts are the c (or number of particles per wafer) chart and the x (or individual measurements) chart. See [Montgomery, 1996]. The basic probability models used for these charts are, respectively, the Poisson distribution and the normal distribution.

In many practical situations, these models may not be appropriate. In such cases, the conclusions drawn from the analysis will be invalid. To solve these problems, we propose to transform data to near normality and then apply the normal-based control charts to the transformed data.

BACKGROUND INFORMATION

Statistical process control techniques have been employed in manufacturing industries to maintain and improve quality by applying statistical methods to the data collected from a process. Control charts are utilized to monitor the process so that out-of-control conditions can be readily observed. Such monitors are effective because they also provide the process engineers with information to identify the causes of the process change and take appropriate action to improve the process. The process may then be adjusted before an excessive number of defective items are produced.

With semiconductor integrated circuit manufacturing entering the sub-half micron era (a half-micron is approximately 20 microinches), smaller critical dimensions (such as the line width and length of a transistor gate) make it increasingly important to closely monitor the process steps by control charting the number of particles on wafers. A wafer is a thin slice of silicon crystal substrate on which several hundred circuits are built. Particles on the wafer surface can lead to failure of the electronic circuit and loss of yield (or percentage of acceptable items). Particles deposited on a wafer surface can be generated by several sources; the manufacturing environment, including people working in the manufacturing area; by the equipment used to manufacture the wafer; and by wafer handling. In particular, equipment-generated particles can be one of the main causes of yield loss. Particle control is a major factor for yield, quality, and reliability. See [Mori, Keeton, and Morrison, 1993]. Therefore, quality improvement of the manufacturing process is of the utmost importance.

Particles generated by the process equipment generally are caused by mechanical operations such as wafer handling, wafer rotation and axis tilt, pumping and venting, and valve operations. In addition, particles can be generated by the chemical and physical reactions of the manufacturing process itself. To minimize particles from these sources, the equipment must be routinely monitored to ensure proper operation. A nonproduction surrogate or test wafer is an unpatterned wafer that has been analyzed by a laser-scanning device and found to contain few or no particles. This wafer is then cycled through the process equipment to simulate the conditions production wafers are exposed to. The wafer is reanalyzed by the laser scanning device and any particles found are categorized by size and location and recorded in a database.

The variable charted will be the number of particles per wafer. The control chart for this variable is usually called the c chart. The Poisson distribution is assumed when using a c chart. In order for the Poisson distribution to apply, three conditions must be satisfied: (a) the defects (or particles) must occur independently of each other, (b) the number of potential locations for defects must be infinitely large, and (c) the probability of occurrence of a defect at any location must be a small constant. In some practical situations, these conditions may not be satisfied. For example, there are particle count data sets, in which the Poisson model is inappropriate.

Another approach would be to use the x or individual measurements chart. In using the x chart, a common and often implicit assumption is that the data are taken from a normal distribution. Unfortunately, this assumption may not be valid. If the data are not from a normal distribution, especially when the underlying distribution is heavily skewed, then the control chart can yield misleading results.

To solve these problems, we transform the original variable to a normal variable. This method has many advantages, notably that the evaluation of the distribution function of the transformed variable is straightforward since tables of the standard normal distribution function and its allied functions are available. In this case study, we will consider several transformations including the Johnson system and the logarithmic and square root transformations. See [Johnson, 1949]. Goodness-of-fit tests, such as chi-square and Shapiro–Wilk, can be used to compare the transformations. See [Shapiro and Wilk, 1965]. The approach is to compute a test score of normality and the associated p-value for each trial transformation. By selecting the highest score or p-value among all trial transformations, one can determine the transformation, along with its parameter estimates, that best fits the data.

Transformations must pass practical criteria too. The selected method must deal reasonably with type I and type II errors—it must detect out-of-control conditions but not send too many false alarms that would desensitize process engineers who monitor charts. The selected method must be easily applied using available tools such as software in the manufacturing environment. The method must be easily understood so that, for example, members who use the method can readily distinguish if an unusual or "outlier" datum is distorting results.

QUESTIONS OF INTEREST

1. If a Poisson or normal model does not adequately describe the particle count data, what can be done about it?
2. One approach to question (a) would be to transform the original random variable to a near-normal variable. How do you find the transformation that best fits the data?
3. How do you find a transformation that can be practically applied?

DATA

Name of Data File: Case08.txt

A piece of new equipment in critical process steps is selected for study. Scanning wafers before and after a process step will reveal the equipment-generated particle counts. The data of equipment-generated particle counts on wafers are recorded and stored in a real-time on-line computer system. A random sample of size 116 is selected from the database. The consecutive numbers in the sample are taken about four days apart. We verify that data are accurate by checking the equipment status log history with the engineers.

Variable Name	Description
particle count	equipment-generated particle count per wafer on a piece of equipment

Notation

X = equipment-generated particle count per wafer
Z = the transformed variable from X

Data

Wafer Number	Particle Count (X)
1	27
2	16
3	16
4	34
5	14
6	13
7	10
.	.
.	.
.	.
.	.
114	19
115	8
116	9

ANALYSIS

The purpose of the analysis is to find out whether the process involving the selected equipment is in control and what areas need improvement. The data set consists of a random sample associated with an unknown distribution. We proceed as follows.

Step 1. Looking at the histogram in Fig. 1, do the data appear to be Poisson or normally distributed?

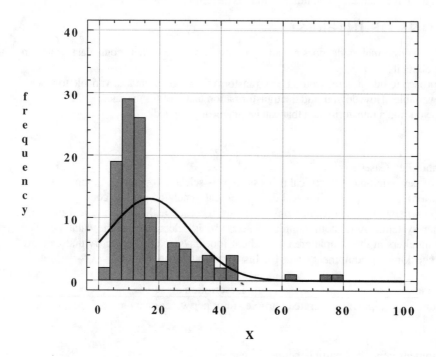

Fig. 1. *Frequency histogram for X data.*

To test that the null hypothesis Ho: X has a Poisson distribution versus the alternative hypothesis Ha: X does not have a Poisson distribution, we use [STATGRAPHICS, 1993]. We choose 21 class intervals with a starting point 0 and class width $80/21 = 3.8095$. A chi-square goodness-of-fit test of Poisson distribution yields the following:

Chi-square = 216.595 with 3 degrees of freedom (d.f.) and p-value < 0.0001.

This means that the data set does not come from a Poisson distribution. In the following discussion, the chi-square tests for normality will also be performed using STATGRAPHICS.

There are many tests for normality. For example, the chi-square test is widely used. The Shapiro–Wilk test has excellent properties compared to other normality tests. See [Shapiro, 1990]. The chi-square test depends on the number and width of class intervals. The Shapiro–Wilk test is more powerful than the chi-square test. Note that the Shapiro–Wilk test is available in statistical software packages such as JMP [Sall and Lehman, 1996] and MINITAB [McKenzie, Schaefer, and Farber, 1995].

For the X data, we test that Ho: X has a normal distribution against Ha: X does not have a normal distribution. We use 21 class intervals with a starting point 0 and class width

80/21 = 3.8095 for the chi-square test. Tests of normality on the particle count data yield the following results:

Chi-square = 82.6951 with 8 d.f. and p-value < 0.00001.

$W = 0.7812$ and p-value < 0.00001, where W is the Shapiro–Wilk test.

Clearly, the data are not from a normal distribution because the p-values are extremely low. The rejection of the normal model leads us to transform the data so that the transformed data may be nearly normally distributed.

Step 2. What transformations are applied?

Three transformations are proposed here which will subsequently be evaluated for their fit to a normal distribution:
1. Johnson system of distributions.
2. Logarithmic transformation.
3. Square-root transformation.

Step 3. The results for the transformed data are as follows:

1. Johnson transformation: See [Chou et al., 1994] and [Chou and Polansky, 1996].
   For the X data, the best-fit Johnson transformation is given by
   $$Z = 2.5350 + 1.1155 \ln((X - 2.2369)/(124.4346 - X)) \text{ for } 2.2369 < X < 124.4346.$$
   For the Z data, we use 21 class intervals with a starting point -4.0 and class width 8/21 = 0.381 for the chi-square test. Tests of normality on Z give the following:
   Chi-square = 8.5882 with 8 d.f. and p-value = 0.378.
   $W = 0.9768$ and p-value = 0.3221.
2. Logarithmic transformation:
   For the X data, the logarithmic transformation is $Z = \ln(X)$. For the Z data, we use 21 class intervals with a starting point 0 and class width 5/21 = 0.2381 for the chi-square test. Tests of normality on Z yield the following:
   Chi-square = 13.0802 with 8 d.f. and p-value = 0.109.
   $W = 0.9711$ and p-value = 0.1251.
3. Square-root transformation:
   For the X data, $Z = (X)^{1/2}$. For the Z data, we use 21 class intervals with a starting point 0 and class width 10/21 = 0.4762 for the chi-square test. Tests of normality on Z yield the following results:
   Chi-square = 24.034 with 8 d.f. and p-value = 0.002.
   $W = 0.9089$ and p-value < 0.0001.

Step 4. Examine the results in Step 3. Do the transformed data appear to be normally distributed?

1. Johnson transformation: Refer to Step 3. We see that the p-values are well above 0.05, and hence we would not reject the hypothesis of a normal model for the transformed data.
2. Logarithmic transformation: Refer to Step 3. We note that the p-values are above 0.05, and hence we would not reject the hypothesis of a normal model for the transformed data.
3. Square-root transformation: Refer to Step 3. We note that the p-values are well below .05, meaning that the normal model does not fit the transformed data well.

Step 5. What is the transformation that best fits the data?

We select the best-fit transformation that corresponds to a maximum p-value among all transformations under consideration. For the X data, the best-fit transformation is the Johnson transformation given in Step 3. The histogram for the transformed data is displayed in Fig. 2.

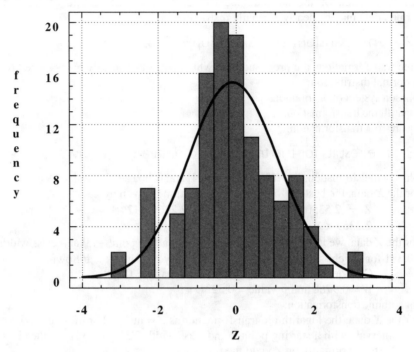

Fig. 2. *Frequency histogram for Johnson transformed Z data.*

Step 6. Control chart for the transformed data:

The transformed data are calculated based on the equation
$$Z = 2.5350 + 1.1155 \ \ln((X - 2.2369)/(124.4346 - X))$$
for $2.2369 < X < 124.4346$.

The moving range (MR) chart for Z has
> center line (CL) = 1.25568,
> upper control limit (UCL) = 4.10230,
> lower control limit (LCL) = 0.

The MR chart for Z is in control and the estimated process standard deviation (or sigma) is 1.11319.

The control chart for Z has
> CL = the average of 116 Z values = −0.09259,
> UCL = CL + 3 sigma = 3.24698,
> LCL = CL − 3 sigma = −3.43216

and is shown in Fig. 3. The Z chart is in control and the estimated process mean is −0.09259.

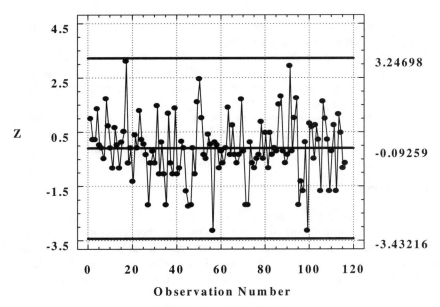

Fig. 3. *Control chart for Z.*

Step 7. Control chart for the raw data:

We can express X in terms of Z as
$$X = 124.4346 - 122.1978/\{\exp[(Z - 2.5350)/1.1155] + 1\}.$$
The control chart for X in Fig. 4 has
  CL = 12.8216 or 13, since particle count data are integers,
  UCL = 82.1991 or 82,
  LCL = 2.8146 or 3.

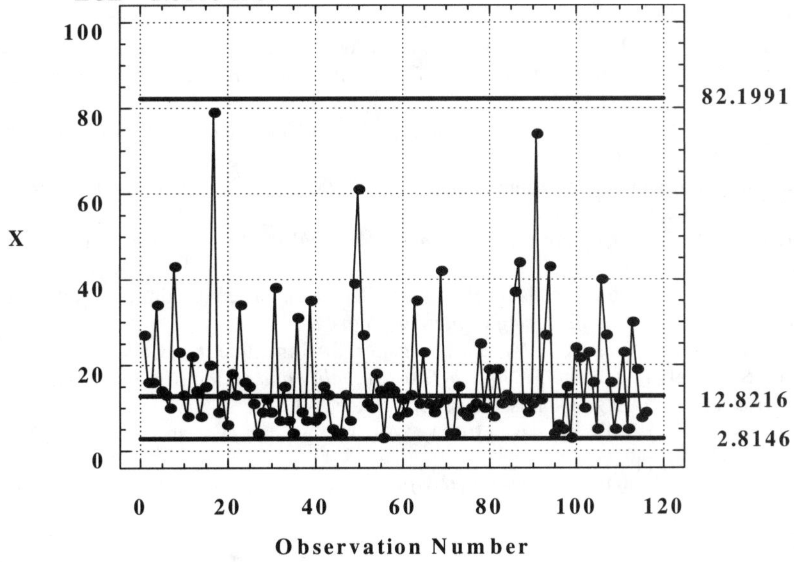

Fig. 4. *Control chart for X.*

Although we could draw conclusions from the control charts at Step 6, we would like to include Step 7 in order to show the following.

Clearly, the control limits for the original variable X are not symmetrical about the center line. This suggests that the 3-sigma control limits should not be applied directly to the nonnormal data. We must transform the nonnormal data to nearly normal and then use the 3-sigma limits for the transformed data.

The Johnson transformation from X to Z is an increasing function. This means that large (or small) values of X correspond to large (or small) Z values. If a point is outside the control limits of the Z chart, it also falls outside the control limits for the X chart. Therefore, it suffices to use the Z chart in Step 6 to draw conclusions.

CONCLUSIONS

There are many useful distributions besides those that are so commonly used, the normal and the Poisson distributions. Conventional Shewhart control charts can be used for process control. However, failure to understand nonnormality leads to over control or under control of the manufacturing process. This discussion demonstrates the various transformations applied to nonnormal data in conjunction with goodness-of-fit tests on whether the data become nearly normal through transformation. The transformation yielding the best fit among the transformations tried will be chosen for the data.

Based on this study, it is clear that with relative ease, quality engineers can produce control charts under nonnormal distributions. It is hoped that this presentation will encourage more practitioners to check the validity of the assumptions for any statistical method before applying it.

REFERENCES

Chou, Y. and Polansky, A. M. (1996), "Fitting SPC Data Using a Sample Quantile Ratio," *ASA Proceedings of the Quality and Productivity Section*. American Statistical Association, Alexandria, VA.

Chou, Y., Turner, S., Henson, S., Meyer, D., and Chen, K. S. (1994), "On Using Percentiles to Fit Data by a Johnson Distribution," *Communications in Statistics - Simulation and Computation*, pp. 341–354.

Johnson, N. L. (1949), "Systems of Frequency Curves Generated by Methods of Translation," *Biometrika*, pp. 149–176.

McKenzie, J., Schaefer, R. L., and Farber, E. (1995), *The Student Edition of MINITAB for Windows*, Addison–Wesley, New York.

Montgomery, D. C. (1996), *An Introduction to Statistical Process Control*. Third Edition, John Wiley & Sons, New York.

Mori, K., Keeton, D., and Morrison, P. (1993), "Yield Enhancement through Particle Control," *SEMI Ultraclean Manufacturing Conference*, pp. 47–51.

Sall, J. and Lehman, A. (1996), *JMP Start Statistics*, SAS Institute Inc., Cary, NC.

Shapiro, S. S. (1990), *Volume 3: How to Test Normality and Other Distributional Assumptions*, Revised edition, ASQC Quality Press, Milwaukee, WI.

Shapiro, S. S. and Wilk, M. B. (1965), "An Analysis of Variance Test for Normality," *Technometrics*, pp. 355–370.

STATGRAPHICS (1993), *STATGRAPHICS Plus Version 7*, Manugistics, Inc., Rockville, MD.

BIOGRAPHIES

Youn-Min Chou is a professor of statistics at the University of Texas at San Antonio. She received a Ph.D. in statistics from Southern Methodist University. Her research interests are in the areas of quality control and process capability. She has served as

Principal Investigator on projects funded by the National Science Foundation and the Texas Higher Education Coordinating Board. In 1996, she received a Faculty Development Leave Award from the University of Texas at San Antonio.

Galen Halverson is the manager of the Manufacturing Science Research Center at the San Antonio manufacturing operation of Sony Semiconductor Company of America. He has a Bachelors in Mathematics and a Masters in Statistics; both degrees are from the University of Wyoming. Galen has diverse experience ranging from Air Force pilot to Nuclear Statistics expert for an agency of the UN. His semiconductor experience includes statistical consulting, managing process engineering, and strategic planning.

Steve Mandraccia received an M.S. in Statistics from the University of Texas at San Antonio. He worked on his thesis as a co-op statistician at Sony Semiconductor Company of America. He has a Bachelors in Computer Science from Regis University in Denver, Colorado. He has extensive experience in the disk drive industry.

CHAPTER 9

EVALUATION OF SOUND TO IMPROVE CUSTOMER VALUE

John R. Voit and Esteban Walker

The rumble of a motor, the tick of a clock, the hum of a generator are all product sounds that are not necessary for product performance. Some of these unnecessary noises may be pleasant to the consumer, whereas others may be annoying or even intolerable. Unfortunately, "pleasantness" is inherently a subjective characteristic and thus is difficult to measure reliably. Subjective characteristics are often evaluated through a panel of judges using a method to rank the items. There is little information on which method provides the most reliable and consistent rankings. This article compares two subjective evaluation methods commonly used by panels of judges to rate noises. Using data from a manufacturing process, the methods are compared on the basis of

- Consistency of judges within panels,
- Consistency of panels over time,
- Agreement between an expert panel and a nonexpert panel.

INTRODUCTION

There exists an increased need in the engineering community to evaluate noise "quality" to justify adding unit cost to better satisfy the consumer. For instance, the sound of a car air conditioning (AC) system is known to be annoying to some customers. This fact has been conveyed through warranty claims for noise complaints and customer clinics, where people were asked about their AC units. The engineers have several design options that will reduce the noise generated by the AC unit; however, all will increase the cost of production.

With a reliable method to evaluate the AC unit noise, engineers and managers would be better able to determine the option that represents the best value to the consumer. The same principles can be applied to other products where costly design changes are considered to reduce objectionable noises.

BACKGROUND INFORMATION

Two procedures have gained widespread use for the evaluation of noise. Both methods rely on a panel of evaluators listening to sounds and registering an opinion. The purpose of this paper is to compare these methods with respect to several criteria. The noises involved in this study were produced by different car AC motors.

The first procedure is the "paired comparison" (PC) method. In this method, each judge listens to pairs of sounds and chooses the one that is more "pleasant." The total number of comparisons for each judge is, therefore, $k(k-1)/2$, where k is the number of sounds. For every judge, the number of marks for each sound is totaled. A 1 is added to each of these totals in order to transform them into ranks. The ranks are then summed across the judges to get a panel score.

The second procedure considered is the "Free Scale" (FS) method. This method consists of listening to a series of sounds and asking each judge to assign an arbitrary score (higher for more pleasant) to the first sound heard and then to assign scores to the other sounds relative to each other. Each sound is repeated p times and the order is randomized. The total number of sounds that each judge hears is pk. To eliminate the effect of scale, the scores for each sound are standardized so that the maximum is 1 and the minimum 0. These standardized scores are averaged for each sound and transformed into midranks to create each judge's ranking. A panel score is obtained by averaging the rankings across judges.

In order to evaluate the consistency of ratings for both methods, two panels of judges were formed. The members of the first "expert" panel were engineers who are familiar with the nature of the sounds and the rating procedures. The second "nonexpert" panel was formed with students of the M. S. program in Statistics at the University of Tennessee. These students were unfamiliar with both the sounds and the rating procedures. The purpose was to create a panel of judges that would be more similar to the average car buyer.

Each panel consisted of seven judges who rated twelve sounds with each method. The procedure was repeated approximately one month later with different randomizations of the sounds. High quality Digital Audio Tapes (DAT) and equipment were used throughout the study. Special care was taken to ensure that the volume was constant throughout the tapes. The same equipment was used for both panels.

QUESTIONS OF INTEREST

The main objective of the study was to determine if one of the methods was superior to the other. The criteria used to compare the methods were
- Consistency of judges within panels,
- Consistency of panels over time,
- Agreement between an expert panel and a nonexpert panel.

For example, does a panel have "an opinion?" That is, is there agreement among the judges? Does a panel have the same opinion throughout time? Do the two panels (expert and nonexpert) have similar opinions?

DATA

Tables 1 and 2 display the number of votes given by each judge to the sounds for each session using the PC method.

Table 1. *Votes assigned by the nonexpert panel using the PC method in the two sessions.*

Sound	Judge													
	1		2		3		4		5		6		7	
	Session		Session		Session		Session		Session		Session		Session	
	1	2	1	2	1	2	1	2	1	2	1	2	1	2
1	5	7	3	2	5	5	6	10	6	6	8	3	3	4
2	10	11	10	4	6	6	9	9	8	11	7	7	9	8
3	5	3	8	6	8	7	5	4	6	7	4	6	7	6
4	9	6	10	6	4	8	7	8	10	9	8	8	5	7
5	9	10	9	5	4	7	11	7	11	9	11	10	11	6
6	7	7	6	9	8	6	8	7	8	6	5	6	7	9
7	9	7	7	6	5	5	5	6	7	7	9	5	5	6
8	5	4	4	10	8	8	7	5	3	4	5	8	9	8
9	2	6	4	5	6	6	5	5	1	2	2	6	5	7
10	4	4	3	9	8	5	2	4	4	4	6	6	4	4
11	1	1	2	4	4	3	1	1	2	1	1	1	1	1
12	0	0	0	0	0	0	0	0	0	0	0	0	0	0

Table 2. *Votes assigned by the expert panel using the PC method in the two sessions.*

Sound	Judge													
	1		2		3		4		5		6		7	
	Session		Session		Session		Session		Session		Session		Session	
	1	2	1	2	1	2	1	2	1	2	1	2	1	2
1	9	5	5	5	11	10	9	10	6	4	8	7	7	8
2	9	10	9	10	8	9	10	10	6	7	9	10	9	9
3	7	7	6	9	7	10	7	8	6	7	6	6	7	7
4	9	8	7	8	9	8	8	7	10	10	9	10	9	8
5	10	11	9	10	10	8	10	8	11	10	10	10	10	11
6	5	6	7	5	4	5	5	5	7	6	7	5	6	6
7	7	8	9	7	6	6	6	8	8	5	6	4	5	6
8	2	4	8	5	5	3	4	2	5	9	4	5	4	2
9	3	2	3	3	3	3	3	4	4	5	3	4	3	5
10	4	4	2	3	2	3	3	3	2	2	3	4	5	3
11	1	1	1	1	1	1	1	1	1	1	1	1	1	1
12	0	0	0	0	0	0	0	0	0	0	0	0	0	0

In order to compare the two methods, the scores from the FS method were transformed into the ranks given in Tables 3 and 4. The standardized scores appear in the Appendix.

Table 3. *Midranks assigned by the nonexpert panel using the FS method in the two sessions (higher ranks correspond to "better" sounds).*

Sound	Judge													
	1		2		3		4		5		6		7	
	Session		Session		Session		Session		Session		Session		Session	
	1	2	1	2	1	2	1	2	1	2	1	2	1	2
1	10	3	10	4	9	8	9.5	12	8.5	5.5	11	3	6	3.5
2	7	9.5	12	4	12	11	11.5	10.5	10.5	8	10	6.5	8.5	10.5
3	3.5	7.5	8	6	6.5	3	6.5	4.5	10.5	12	7.5	6.5	7	7
4	10	12	9	7.5	6.5	9.5	6.5	9	6	10.5	6	12	4.5	9
5	12	9.5	11	1.5	10.5	12	11.5	10.5	12	10.5	12	10.5	12	12
6	8	5	4.5	7.5	3	9.5	6.5	7	6	8	7.5	6.5	8.5	7
7	10	5	7	12	10.5	6.5	9.5	7	6	8	4	6.5	10.5	3.5
8	3.5	11	3	10.5	4	6.5	6.5	4.5	3	3.5	4	10.5	10.5	10.5
9	5.5	7.5	2	10.5	6.5	4.5	3	7	4	3.5	9	6.5	4.5	7
10	5.5	5	4.5	9	6.5	4.5	4	3	8.5	5.5	4	6.5	3	5
11	2	2	6	1.5	2	2	2	2	2	2	2	2	2	1.5
12	1	1	1	4	1	1	1	1	1	1	1	1	1	1.5

Table 4. *Midranks assigned by the expert panel using the FS method in the two sessions (higher ranks correspond to "better" sounds).*

Sound	Judge													
	1		2		3		4		5		6		7	
	Session		Session		Session		Session		Session		Session		Session	
	1	2	1	2	1	2	1	2	1	2	1	2	1	2
1	10.5	10.5	8	5	12	12	7	8	12	8	10.5	10	8.5	12
2	9	8	10.5	10	11	11	11	9.5	5.5	8	8	11.5	11	10.5
3	6	10.5	10.5	12	9	9.5	9	12	10	3	7	5	8.5	8.5
4	10.5	10.5	10.5	11	10	7.5	12	7	9	10.5	9	11.5	12	8.5
5	12	10.5	10.5	9	8	9.5	10	9.5	11	8	12	8	10	10.5
6	6	4.5	4.5	8	7	6	5	5	8	12	5.5	3	6	6
7	6	7	7	6	6	7.5	8	11	7	5	10.5	9	7	7
8	6	4.5	4.5	3	5	5	4	2	3	5	5.5	7	5	5
9	3	4.5	4.5	4	4	3.5	3	3	4	10.5	4	6	3	4
10	6	4.5	4.5	7	3	3.5	6	6	5.5	5	3	4	4	3
11	2	1.5	2	2	1.5	2	2	4	2	2	2	2	1.5	2
12	1	1.5	1	1	1.5	1	1	1	1	1	1	1	1.5	1

ANALYSIS

Panel scores were obtained for each session of both the FS and PC method. These scores were calculated by summing ranks across all evaluators (within a panel) for each sound (Tables 5 and 6). To transform the votes for individual judges into ranks (from 1 to 12), the scores for the PC method were obtained by adding seven to the total number of votes.

Table 5. *PC scores for each panel (N = nonexpert, E = expert) and each session (1, 2).*

Rank	Sound ID	PCN1	Sound ID	PCN2	Sound ID	PCE1	Sound ID	PCE2
1	5	73	2	63	5	77	5	75
2	2	66	5	61	4	68	2	72
3	4	60	4	59	2	67	4	66
4	6	56	6	57	1	62	3	61
5	7	54	8	54	7	54	1	56
6	3	50	7	49	3	53	7	51
7	8	48	3	46	6	48	6	45
8	1	43	1	44	8	39	8	37
9	10	38	9	44	9	29	9	33
10	9	32	10	43	10	28	10	29
11	11	19	11	19	11	14	11	14
12	12	7	12	7	12	7	12	7

Table 6. *FS scores for each panel (N = nonexpert, E = expert) and each session (1, 2).*

Rank	Sound ID	FSN1	Sound ID	FSN2	Sound ID	FSE1	Sound ID	FSE2
1	5	81	4	69.5	5	73.5	2	68.5
2	2	71.5	5	66.5	4	73	4	66.5
3	1	64	2	60	1	68.5	1	65.5
4	7	57.5	8	57	2	66	5	65
5	3	49.5	6	50.5	3	60	3	60.5
6	4	48.5	7	48.5	7	51.5	7	52.5
7	6	44	3	46.5	6	42	6	44.5
8	10	36	9	46.5	8	33	9	35.5
9	8	34.5	1	39	10	32	10	33
10	9	34.5	10	38.5	9	25.5	8	31.5
11	11	18	11	13	11	13	11	15.5
12	12	7	12	10.5	12	8	12	7.5

The scores from the previous tables were plotted to investigate the agreement between the panels for each evaluation method (Figures 1 and 2). It was noted that sounds 11 and 12 (shown by "x") scored in the bottom two positions, regardless of evaluation method, panel, or session. These sounds were eliminated from future calculations because the engineering objective was to understand the differences between the evaluation methods when the sounds were difficult to discern, not when the sounds were clearly unpleasant. From a statistical point of view, these sounds were highly influential, and it was considered that more truthful measures of consistency and agreement were obtained by eliminating these influential points.

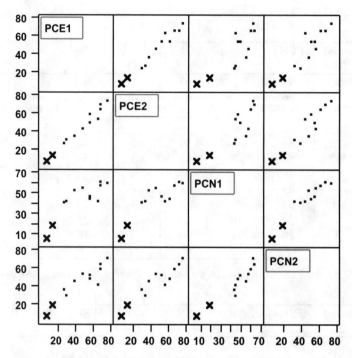

Fig. 1. *Scatterplots of PC scores between panels and across time.*

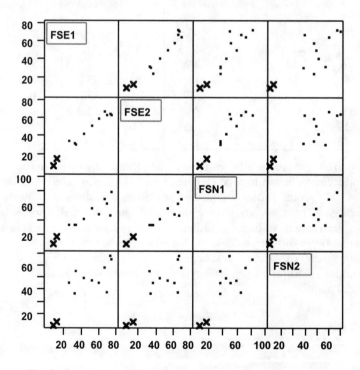

Fig. 2. *Scatterplots of FS scores between panels and across time.*

Kendall's *W* coefficient of concordance was used to measure the internal consistency (agreement) among judges' rankings within a panel. The coefficient is defined as

$$W = \frac{12}{n^2 k(k^2-1)} \sum_{j=1}^{k} (R_j - n(k-1)/2)^2,$$

where R_j is the sum of the ranks for the *j*th sound, *n* is the number of judges, and *k* is the number of sounds tested. Another form of this coefficient that sheds some light on its meaning (see [Hettmansperger, 1984, p. 210]) is

$$W = \frac{n-1}{n} r_{ave} + \frac{1}{n},$$

where r_{ave} is the average of the Spearman correlation between the ranks assigned by each pair of judges (see Tables 1 and 2).

It can be shown that Kendall's *W* varies between 0, when concordance is null, and 1, when concordance is perfect among the judges. Null concordance is interpreted to mean that the judges assign ranks at random to the sounds. As *W* increases, the opinion of the panel members is more consistent. The hypothesis of no concordance (opinion) within a panel can be tested using a Chi-square test. Under the null hypothesis of null concordance, $W \cdot (n(k-1))$ is distributed as a Chi-square with $k-1$ degrees of freedom (see [Hettmansperger, 1984, p. 211]).

The values of *W* for the PC method appear on the diagonal of Table 7. For the expert panel the values were .7831 and .7114 for sessions 1 and 2, respectively. The corresponding values for the nonexpert panel were .3880 and .2875. All these values were significant at the .05 level, indicating that both panels had internal consistency. However, the expert panel was considerably more internally consistent than the nonexpert panel.

To evaluate agreement between panels for the PC method, the Spearman correlation of the scores (Tables 5 and 6) was computed between each pair of panels. These values appear in the off-diagonal of Table 7. The results suggest significant agreement between the panels and sessions ($p < .05$).

Table 7. *Concordance within and between panels using the PC method without sounds* 11 *and* 12. *P values in parentheses.*

	Panel	(Session)		
	Expert (1)	Expert (2)	Nonexpert (1)	Nonexpert (2)
Expert (1)	.7831 (<.0001)			
Expert (2)	.9848 (<.0001)	.7114 (<.0001)		
Nonexpert (1)	.7195 (.0190)	.7477 (.0129)	.3880 (.0037)	
Nonexpert (2)	.8146 (.0041)	.8424 (.0022)	.9301 (<.0001)	.2875 (.0339)

The same procedure was used to evaluate consistency for the FS method (Tables 3 and 4). The results appear in Table 8. As before, the internal consistency (diagonal values) was significant in all instances, with higher concordance associated with the expert panel. However, none of the concordance values involving the nonexpert panel in session 2 were significant at the .05 level. This result indicates that, while the nonexpert panel in session 2 was consistent internally, its ranks did not agree with the rest of the panels.

Table 8. *Concordance within and between panels using the FS method without sounds* 11 *and* 12. *P values in parentheses.*

	Panel	(Session)		
	Expert (1)	Expert (2)	Nonexpert (1)	Nonexpert (2)
Expert (1)	.7727 (<.0001)			
Expert (2)	.8424 (.0022)	.6183 (<.0001)		
Nonexpert (1)	.8328 (0028)	.7964 (.0058)	.5745 (<.0001)	
Nonexpert (2)	.5593 (.0928)	.4377 (.2058)	.2957 (.4068)	.2870 (.0342)

During the analyses, it was noted that in some sessions, some judges dissented considerably from the rest. The scores of these judges had a sizable impact on the measures of concordance. For example, by eliminating one nonexpert judge from the second session, the internal consistency measured by Kendall's W increased from .2875 to .4111 and from .2870 to .4802 for the PC and FS methods, respectively.

CONCLUSIONS

This investigation compared two subjective noise evaluation methods and two panels of judges. The PC method is significantly easier to analyze since the votes can be used directly as ranks, whereas the FS method calls for standardization of the scores (see Appendix Tables A1 and A2). Analyses using nonparametric statistics based on ranks yielded the following:

- The expert panel was more internally consistent than the nonexpert panel, regardless of the evaluation method.
- For both methods, the expert panel was consistent over time. However, the nonexpert panel was only consistent over time using the PC method.
- There was agreement between the two panels for the PC method. The same cannot be stated for the FS method.
- Clearly inferior (or superior) sounds can have a strong incremental and misleading effect on the measures of agreement.
- A single dissenting judge can notably affect the panel's internal consistency and presumably the level of agreement between panels.

Based on these findings, if nonexpert judges are involved in a panel evaluation, the PC method should be used to increase the internal consistency and consistency over time. A decision should be made on whether to include dissenting judges and clearly different items in the analysis, since these can have a considerable effect on the level of concordance. For example, a hearing deficit is a valid reason to eliminate a judge's scores.

The ultimate validation of subjective scores would be their correlation with objective noise measures like amplitude, frequency, etc. If these correlations can be established, then objective measures of noise quality could be devised.

REFERENCE

Hettmansperger, T. P. (1984), *Statistical Inference Based On Ranks*, Wiley, New York.

APPENDIX

Table A1. *Standardized scores for the nonexpert panel using the FS method in the two sessions.*

Sound	Judge												
	1		2		3		4		5		6		7	
	Session		Session		Session		Session		Session		Session		Session	
	1	2	1	2	1	2	1	2	1	2	1	2	1	2
1	0.71	0.67	0.56	0.06	0.79	0.71	0.80	0.87	0.83	0.56	0.81	0.62	0.63	0.50
2	0.64	0.81	0.69	0.06	0.92	0.88	0.93	0.73	0.92	0.67	0.76	0.67	0.73	0.83
3	0.52	0.76	0.43	0.11	0.75	0.54	0.73	0.53	0.92	0.78	0.67	0.67	0.67	0.67
4	0.71	0.95	0.49	0.17	0.75	0.75	0.73	0.67	0.75	0.72	0.62	0.81	0.60	0.75
5	0.76	0.81	0.57	0.00	0.83	0.92	0.93	0.73	1.00	0.72	0.95	0.76	0.83	0.92
6	0.67	0.71	0.27	0.17	0.62	0.75	0.73	0.60	0.75	0.67	0.67	0.67	0.73	0.67
7	0.71	0.71	0.36	0.44	0.83	0.67	0.80	0.60	0.75	0.67	0.57	0.67	0.77	0.50
8	0.52	0.90	0.20	0.33	0.71	0.67	0.73	0.53	0.42	0.44	0.57	0.76	0.77	0.83
9	0.62	0.76	0.09	0.33	0.75	0.62	0.60	0.60	0.50	0.44	0.71	0.67	0.60	0.67
10	0.62	0.71	0.27	0.22	0.75	0.62	0.67	0.33	0.83	0.56	0.57	0.67	0.50	0.58
11	0.14	0.29	0.33	0.00	0.33	0.29	0.27	0.13	0.25	0.11	0.24	0.24	0.17	0.17
12	0.10	0.10	0.00	0.06	0.21	0.04	0.00	0.00	0.00	0.00	0.05	0.10	0.03	0.17

Table A2. *Standardized scores for the expert panel using the FS method in the two sessions.*

Sound	Judge												
	1		2		3		4		5		6		7	
	Session		Session		Session		Session		Session		Session		Session	
	1	2	1	2	1	2	1	2	1	2	1	2	1	2
1	0.94	1.00	0.87	0.43	0.93	1.00	0.70	0.77	1.00	0.67	0.87	0.84	0.80	1.00
2	0.89	0.83	0.93	0.84	0.82	0.85	0.88	0.80	0.75	0.67	0.82	0.92	0.88	0.75
3	0.83	1.00	0.93	0.96	0.69	0.62	0.82	0.85	0.89	0.39	0.71	0.40	0.80	0.67
4	0.94	1.00	0.93	0.86	0.76	0.55	1.00	0.73	0.87	0.72	0.85	0.92	0.93	0.67
5	1.00	1.00	0.93	0.81	0.63	0.62	0.87	0.80	0.95	0.67	0.93	0.73	0.82	0.75
6	0.83	0.50	0.67	0.74	0.58	0.38	0.56	0.39	0.80	0.89	0.70	0.36	0.58	0.50
7	0.83	0.67	0.80	0.64	0.44	0.55	0.78	0.83	0.77	0.56	0.87	0.77	0.68	0.58
8	0.83	0.50	0.67	0.38	0.38	0.32	0.47	0.16	0.69	0.56	0.70	0.49	0.57	0.42
9	0.75	0.50	0.67	0.41	0.31	0.23	0.39	0.33	0.71	0.72	0.69	0.44	0.31	0.33
10	0.83	0.50	0.67	0.73	0.26	0.23	0.58	0.48	0.75	0.56	0.60	0.37	0.38	0.25
11	0.19	0.00	0.27	0.17	0.00	0.08	0.26	0.38	0.07	0.28	0.18	0.12	0.00	0.08
12	0.17	0.00	0.00	0.00	0.00	0.00	0.00	0.00	0.00	0.00	0.00	0.00	0.00	0.00

BIOGRAPHIES

John R. Voit works for Delphi Harrison Thermal Systems, where he applies statistics to product engineering. Delphi is a leader in the automotive engine cooling and heating, ventilation, and air conditioning systems. Mr. Voit has an M.S. in Statistics from the University of Tennessee.

Esteban Walker is an Associate Professor in the Department of Statistics at the University of Tennessee, Knoxville. His main interest is in the application of statistics to industrial problems. Dr. Walker obtained a doctorate in Statistics from Virginia Tech.

CHAPTER 10

IMPROVING INTEGRATED CIRCUIT MANUFACTURE USING A DESIGNED EXPERIMENT

Veronica Czitrom, John Sniegowski, and Larry D. Haugh

Integrated circuits (chips) are essential components in the electronics industry, where they are used in computer products, radios, airplanes, and other electronic equipment. Numerous integrated circuits are manufactured simultaneously on one silicon wafer. Chemical etching (removing an oxide layer from a wafer) is one of many manufacturing steps in the creation of an integrated circuit. A designed experiment was performed to improve control of an etching process. It was necessary to increase the CF_4 gas flow beyond what development engineers had recommended, and it was hoped that two other factors, electric power and bulk gas flow, could be used to offset the effect of this increase on three important responses related to yield and throughput: etch rate, etch rate nonuniformity, and selectivity. The designed experiment allowed a systematic and efficient study of the effect of the three factors on the responses. Settings were found that allowed the CF_4 gas flow to be increased.

INTRODUCTION

The semiconductor industry is the foundation of the electronics industry, the largest industry in the U.S., employing 2.7 million Americans. The semiconductor industry manufactures integrated circuits, or chips, for use in airplanes, computers, cars, televisions, and other electronic equipment. Each integrated circuit consists of thousands of interconnected microscopic elements such as transistors and resistors. The smallest active features are 0.5 microns in width, or approximately 1/150th the diameter of a human hair. During manufacture, many integrated circuits are created simultaneously on a thin round silicon wafer. The wafer goes through a very complex set of manufacturing steps that can take up to two months to complete. The steps include depositing layers of material on the wafer, creating patterns on the wafer through a photolithographic process, etching away part of the material, cleaning in chemical baths, and implanting dopant atoms to change the electrical properties of the silicon substrate.

During the creation of a new integrated circuit technology, a new process was developed by the research and development group to etch silicon dioxide (oxide) from wafers using a CF_4 gas. This process was then transferred to manufacturing. The process development group had suggested using a CF_4 gas flow rate of 5 sccm (standard cubic centimeters per minute). However, the manufacturing equipment that was to be used for the etching process is under better control if the CF_4 gas flow is at least 15 sccm. For this reason, the manufacturing engineer wanted to increase the flow of CF_4 gas from 5 sccm to 15 sccm. He thought adjustments to two other processing factors, power and bulk gas flow, might offset the effect of the increase in CF_4 gas flow on two important responses, wafer throughput and yield.

A designed experiment is an efficient, methodical approach to discern and quantify the effect of controllable process factors on one or more response variables in order to find good processing conditions. For this reason, a designed experiment was used to study the effect of CF_4 gas flow, power, and bulk gas flow on wafer throughput and on yield. A full factorial designed experiment in eight experimental runs, using all possible combinations of two levels of each of the three process factors, as well as two center point experimental runs at the mid-levels (center point) of the three process factors, was performed. Analysis of the experimental results using graphical analysis and multiple regression analysis allowed the discovery of processing conditions in power and bulk flow that permitted the increase in CF_4 flow without having a negative impact on throughput or yield. These results were later confirmed, and the new processing conditions were adopted for manufacturing.

BACKGROUND INFORMATION

Figure 1 shows a cross section of one of millions of transistor gate structures of one of the integrated circuits on a silicon wafer. Figures 1(a) and 1(b) illustrate the structure before and after etching an oxide (silicon dioxide, SiO_2) layer to leave "spacers" next to the polysilicon gate structure. The spacers help set the width of the transistor.

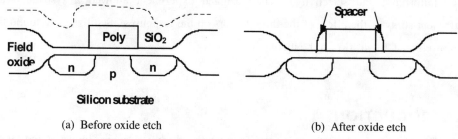

(a) Before oxide etch   (b) After oxide etch

Fig. 1. *Cross section of a transistor before and after oxide etch.*

The etching process is illustrated in Figure 2. During the etching process, a single wafer is introduced into a chamber, placed on a cathode, and a radio frequency power differential is applied. CF_4 gas is introduced into the chamber to etch the oxide for 30 seconds. The CF_4 gas is a trace element in the bulk, or carrier, gas flow of Argon and CHF_3 (trifluoromethane, or Freon-23).

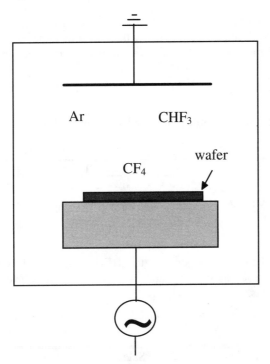

Fig. 2. *Oxide etching process.*

The research and development group that had developed the process had suggested using a CF_4 gas flow rate of 5 sccm for the oxide etching process. However, the manufacturing engineer wanted to increase the CF_4 gas flow to 15 sccm to make the equipment used for manufacture more controllable. He thought that the effect of the increase in CF_4 gas flow on throughput and yield could be offset by changing the settings of applied power and bulk gas flow. The effect on wafer throughput was measured using oxide etch rate, and the effect on yield was measured using etch rate nonuniformity and selectivity.

Etch rate is the amount of oxide that is removed from the wafer per unit time. In this process, etch rate values are around 3,000 Ångstroms/minute. To put this number in perspective, note that there are 10^{10} Ångstroms in one meter (about a yard), and that an atom is several Ångstroms in diameter. The higher the etch rate, the higher the manufacturing throughput. *Etch rate nonuniformity* is a measure of the variability (standard deviation) in oxide etch rate on a wafer expressed as a percentage of the mean oxide etch rate. Since there are integrated circuits throughout the wafer surface, and each integrated circuit needs to function properly, it is important to reduce the variability in the amount of oxide etched at different sites across the wafer. The lower the etch rate nonuniformity the higher the yield, which is the proportion of integrated circuits that function properly at the end of the manufacturing process. *Selectivity* is the ratio of the rate at which oxide is etched to the rate at which polysilicon is etched. As the oxide layer is being etched, it is important to minimize the loss of the polysilicon structure. If oxide is etched at a faster rate than polysilicon, then selectivity is high, and yield is high.

A designed experiment was performed to study the effect of three manufacturing process factors—bulk flow, CF_4 flow, and power—on the three responses—selectivity, etch rate, and etch rate nonuniformity. A statistical model for each response was used to study the effect of the three factors on the response. The models were used to determine whether there are processing conditions that allow the CF_4 flow to be increased to 15 sccm without having a negative impact on the responses. In particular, selectivity should not be degraded.

DATA

A designed experiment was performed in three factors at two levels each:

Factor	Levels	
	Low	High
Bulk gas flow (sccm)	60	180
CF_4 Flow (sccm)	5	15
Power (Watts)	550	700

In the bulk gas flow, the ratio of Argon flow to CHF_3 gas flow was held constant at a 2-to-1 ratio. Other factors such as pressure and magnetic field were held constant at the current operating conditions.

Table 1. *Experimental design in three factors with three responses.*

Run	Bulk gas flow (sccm)	CF_4 flow (sccm)	Power (Watts)	Selectivity	Etch rate (Å/min)	Etch rate nonUniformity
1	60	5	550	10.93	**2710**	11.7
2	180	5	550	**19.61**	2903	**13.0**
3	60	15	550	7.17	3021	9.0
4	180	15	550	12.46	3029	10.3
5	60	5	700	10.19	3233	10.8
6	180	5	700	17.5	3679	12.2
7	60	15	700	**6.94**	3638	**8.1**
8	180	15	700	11.77	**3814**	9.3
9	120	10	625	11.61	3378	10.3
10	120	10	625	11.17	3295	11.1

The experimental design is given in Table 1. The design is a 2^3 full factorial in the eight possible combinations of settings of the three factors at two levels each (runs 1 to 8), with two centerpoints (runs 9 and 10). Centerpoints are design conditions such that each factor is set at levels midway between the low and high settings for that factor. The order of the full factorial portion of the experiment (runs 1 to 8) was randomized. One centerpoint was performed at the beginning of the experiment, and the other centerpoint was performed at the end, to check for changes over time.

Four wafers were etched at each one of the ten treatment combinations (runs) in the experimental design. The first two wafers were dummy wafers, to allow the etching reactor to warm up. The third wafer was an oxide wafer with a layer of silicon dioxide on a bare silicon wafer, which was used to measure the oxide etch rate and the oxide etch rate nonuniformity. The fourth wafer was a poly (polysilicon) wafer with a layer of polycrystalline silicon and a layer of silicon dioxide above the silicon wafer, which was used to measure the polysilicon etch rate in order to evaluate selectivity as the ratio of the oxide etch rate to the polysilicon etch rate. There were only enough wafers available to replicate the centerpoint.

The sampling plan, shown in Figure 3, consisted of taking thickness measurements at 49 sites on each wafer. Oxide thickness was measured at the same 49 sites on each wafer before and after etch. The difference in oxide thickness before and after etch was evaluated at the 49 sites, and the etch rate at each site was calculated as the difference in oxide thickness divided by the etching time (half a minute). The etch rate given in Table 1 is the average of the 49 oxide etch rates on each wafer. The etch rate nonuniformity (often called etch rate *uniformity*) given in Table 1 is the coefficient of variation, calculated by dividing the standard deviation of the 49 oxide etch rates on a wafer by the average of the 49 oxide etch rates on the wafer and multiplying by 100.   The etch rate nonuniformity is the variability in oxide etch rates across the wafer expressed as a percentage of the mean. The polysilicon etch rate is calculated just like the oxide etch rate, using the poly wafer instead of the oxide wafer. Selectivity is the quotient of the average oxide etch rate at the 49 sites on the oxide wafer divided by the average polysilicon etch rate at the 49 sites on the polysilicon wafer.

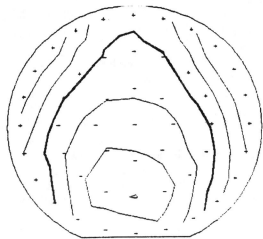

Fig. 3. *Sampling plan of 49 sites on a 6" oxide wafer and contour plots of raw thickness data.*

The highest and lowest values of the three responses are highlighted in Table 1. The engineer looked at these values, as well as the raw data illustrated for a typical wafer in Figure 3, to check for unusual values such as outliers (unusually large or small values) or patterns that could have a strong influence on the analysis or could give additional insight into the process.

QUESTIONS OF INTEREST

The statistical and engineering questions of interest concern the effect of the three experimental factors on the responses. How does each factor affect each response? What graphs would be helpful in identifying recommended process conditions to optimize the process? What graphs would aid in understanding possible factor interactions? What multiple regression equations can be identified and fit for each relationship? Can the critical goal of increasing the CF_4 flow to 15 sccm be achieved without degrading yield and selectivity?

DATA

The data consists of ten experimental runs in three factors with three responses. The three factors are bulk gas flow, CF_4 flow, and power. The three responses are selectivity, oxide etch rate, and oxide etch rate nonuniformity. The first three runs are as follows:

Run	Bulk gas flow (sccm)	CF_4 flow (sccm)	Power (Watts)	Selectivity	Etch rate (Å/min)	Etch rate nonuniformity
1	60	5	550	10.93	2710	11.7
2	180	5	550	19.61	2903	13.0
3	60	15	550	7.17	3021	9.0

ANALYSIS

1. Graphical analysis of the selectivity response.
A. Draw the three main effects plots for selectivity, each of which goes from the average of the four responses at the low level of the factor to the average of the four responses at the high level of the factor.
B. Does selectivity increase or decrease as each factor increases? By how much does selectivity increase/decrease with each factor?
C. Which factor has the greatest effect on selectivity? Which factor has the least effect on selectivity?
D. Add the centerpoints to the main effects plots. Do the centerpoints appear to indicate the presence of curvature in the relationship of selectivity with the three factors?
E. Make the interaction graph of bulk flow and CF_4 flow, with one line for low bulk flow going from the average of the two selectivities at low CF_4 flow to the average of the two selectivities at high CF_4 flow, and the other line for high bulk flow going from the average of the two selectivities at low CF_4 flow to the average of the two selectivities at high CF_4 flow.
F. Make another interaction graph reversing the roles of CF_4 flow and bulk flow in part E.
G. Do the two interaction graphs in parts E and F look the same? For a fixed value of bulk flow, what does each graph indicate about the behavior of selectivity as CF_4 flow increases? For a fixed value of CF_4 flow, what does each graph indicate about the behavior of selectivity as bulk flow increases? Can you draw the same conclusions from both graphs with regard to the behavior of selectivity as a function of bulk flow and CF_4 flow?
H. Make the interaction plots for the other two interactions. Do any of the interactions appear to be important (watch for the extent of nonparallel lines in the interaction graph)?
2. Numerical analysis of the selectivity response—initial model.
A. To compare the effects of the three factors on selectivity using a common scale, code the low and high values of each factor as −1 and +1, respectively.
B. Derive the regression model for selectivity with a constant term, three main effects terms, and three two-factor interaction terms.
C. Derive the corresponding analysis of variance (ANOVA) table.
D. What is the value of R^2, the percentage of the variability explained by the model? Does the model fit the data well?
E. Which is the most significant (important) effect? Which is the least significant effect?

F.   Do the numerical results (size of the coefficients in the model and statistical significance of the terms in the ANOVA table) agree with the conclusions from the main effects plots and the interaction plots in question 1 above?

3.   Numerical analysis of the selectivity response—refined model.

A.   Remove the two least significant terms from the model in question 2 and find the corresponding regression model and ANOVA table.

B.   How does the value of R^2 for this model compare to the value of R^2 for the model with all main effects and interactions in question 2? Did the values of the coefficients change? Did the standard error of the coefficients and their statistical significance change? Why?

C.   Does the significance of the largest interaction term in the model agree with the perceived importance of the interaction as seen in the interaction graph in question 1? If there is a discrepancy, what is its cause?

D.   Derive 95% confidence intervals for the coefficients. Which factor(s) effects are positive, negative, and not significantly different from zero? Does this agree with the graphical analysis of question 1?

4.   Numerical analysis of the selectivity response—contour plot.

A.   Use the fitted model of selectivity derived in question 3 to do a contour plot of selectivity as a function of bulk flow and power, with CF_4 flow fixed at the desired value of 15 sccm. Use the contour plot to answer parts B to F.

B.   For a given value of power, does selectivity increase or decrease as a function of bulk flow?

C.   For a given value of bulk flow, does selectivity increase or decrease as a function of power?

D.   Do the results of parts B and C agree with the conclusions derived from the main effects graphs in question 1?

E.   What do the parallel contour lines indicate about the interaction between bulk flow and CF_4 flow?

F.   What is the maximum value of selectivity in the region? What are the values of power and bulk flow at the point of maximum selectivity?

5.   Do a similar analysis for the etch rate response.

A.   Find the model for etch rate with main effect and interaction terms, and the value of R^2.

B.   Plot the studentized residuals by run number. Is there anything unusual about this graph? Remove run 9 from the remainder of the analysis of etch rate.

C.   Remove the least significant term from the model, and derive the new regression model. Are the coefficients in this new model the same as the coefficients for the initial model with all main effects and interactions? Why?

D.   For a CF_4 flow of 15 sccm, do the contour plot of etch rate as a function of bulk flow and power. Use the contour plot to answer parts E and F.

E.   Why aren't the lines of the contour plot parallel?

F.   Why are the contour lines almost parallel to one of the axes?

G.   What is the highest etch rate in the region? What are the values of power and bulk flow at the point of highest etch rate? Is etch rate much smaller at the low level of bulk flow?

6.   Do a similar analysis for the etch rate nonuniformity response.

A.   Find the model for etch rate nonuniformity in terms of the three main effects and the three two-factor interactions. What is the value of R^2?

B.   Remove the three least significant terms from the regression model.

C.   For a CF_4 flow of 15 sccm, do the contour plot of etch rate nonuniformity as a function of bulk flow and power.

D.   From the contour plot in part C, what is the lowest etch rate nonuniformity in the region? What are the values of power and bulk flow at that point?

7. Analyze the three responses simultaneously using a contour plot.
A. Overlay the contour plots of the three responses (questions 4, 5, and 6) on one graph, for a fixed CF_4 flow of 15 sccm and with power and bulk gas flow along the axes.
B. For what values of power and bulk gas flow is selectivity maximized? What are the values of etch rate and etch rate nonuniformity at that point?
C. Is it possible to simultaneously maximize selectivity and etch rate and minimize etch rate nonuniformity?
8. Prediction using the models for the responses.
A. Use the models for the three responses to predict the responses at the original process conditions of CF_4 flow at 5 sccm, power at 625 Watts, and bulk flow at 90 sccm.
B. Use the three models to predict the responses at the values of the factors that maximize selectivity.
C. Are the values of the three responses very different at the original process conditions (section A) and at the point that maximizes selectivity (section B)?
9. Conclusions.
A. The most important manufacturing question was whether the CF_4 gas flow could be increased from 5 to 15 sccm without having a negative impact on yield and on throughput, as measured by the three responses. Based on this experiment, would you recommend the increase in CF_4 flow to manufacturing management?

PRESENTATION OF RESULTS TO CUSTOMERS

Write a brief report to the project manager summarizing your recommendations. Make sure the summary can be understood by nonstatisticians and that it relates clearly to the engineering problem (feasibility of increasing CF_4 flow). Include detailed statistical analysis issues in an appendix.

Prepare a brief oral presentation of your results for the project manager. Concentrate on a simple graphical presentation of the main results that can be easily understood by a nonstatistician (remember—a picture is worth a thousand words). State the conclusions in terms of the engineering problem. Be prepared to answer questions and to present relevant backup information in case it is needed.

REFERENCES

Box, George E.P, William G. Hunter, and J. Stuart Hunter (1978), *Statistics for Experimenters*. New York, NY: John Wiley and Sons, Inc.

Mason, Robert L., Richard F. Gunst, and James L. Hess (1989), *Statistical Design and Analysis of Experiments with Applications to Engineering and Science*. New York, NY: John Wiley and Sons, Inc.

Douglas C. Montgomery (1997), *Design and Analysis of Experiments* (4th edition), New York, NY: John Wiley and Sons, Inc.

BIOGRAPHIES

Veronica Czitrom is a Distinguished Member of Technical Staff at Bell Labs, Lucent Technologies. She leads consulting and training efforts for the application of statistical methods, primarily design of experiments. Czitrom has published research papers on design of experiments and four engineering textbooks. She received a B.A. (Physics) and an M.S. (Engineering) from the University of California at Berkeley, and a Ph.D. (Mathematics - Statistics) from the University of Texas at Austin. She was a faculty member at The University of Texas at San Antonio and at the Universidad Nacional Autónoma de México.

John Sniegowski has been a plasma process engineer with Lucent Technologies in Orlando, Florida since 1991. Prior to that he spent seven years as a process engineer with Harris Semiconductor. His formal education includes a B.S. in Chemical Engineering from the University of Florida, an MBA from Embry-Riddle, and graduate level statistics courses in Design of Experiments and Statistical Process Control.

Larry Haugh is Professor and Director of the Statistics Program at the University of Vermont. He has had short-term positions with IBM, Shell Research, Oscar Mayer, and the Crane Naval Ammunition Depot. He provided consulting and training for other organizations and collaborated in a variety of medical research studies (particularly for orthopedics and low back pain disability). His education includes a B.A. (Mathematics) from Wabash College and an M.A. (Mathematics), M.S. (Statistics), and Ph.D. (Statistics) from the University of Wisconsin-Madison.

CHAPTER 11

EVALUATING THE EFFECTS OF NONRESPONSE AND THE NUMBER OF RESPONSE LEVELS ON SURVEY SAMPLES

Robert K. Smidt and Robert Tortora

The purpose of this case study is to examine two elements in survey sampling that affect estimation: nonresponse and the number of response levels allowed for any question. The first factor, nonresponse, causes difficulties in surveys, especially when those who fail to respond are liable to be different from those who do. Estimates based on such responses will have substantial errors. The second factor is associated with surveys that employ the Likert scale. The Likert scale offers a series of "k" ordered responses indicating the degree of agreement (or satisfaction or support, etc.) for the question under consideration. The choice of k, the number of available responses to a survey question, is crucial, particularly when interest lies in estimating the percent that belongs in the top category. We would like to examine the combined effect of these two factors on estimation.

INTRODUCTION

Survey samples are used to gather information from many and diverse groups. Based on these, elections are predicted before the polls close, television networks decide which programs to replace, advertising firms choose the group to target with their marketing strategies, and companies retool their factories. It is crucial for sample surveys to be designed so that representative information is obtained from the appropriate group. Failure to do so can lead to disastrous results. Introductory statistics texts enjoy describing the *Literary Digest's* attempt to forecast the Roosevelt/Landon election (remember President Alf?) or presenting the photograph of Harry S. Truman holding aloft the headline proclaiming Dewey's victory. Less dramatic but often more costly mistakes are made when inaccurate sample surveys lead firms to make bad decisions and take inappropriate actions.

The sample survey methodologist tries to produce a questionnaire and sampling procedure that reduce the possibility of a misleading survey. Questions are carefully examined to make sure that they are precise and not misleading and are posed so that they are unlikely to produce biased answers. Each question is evaluated to see if it is likely to lead to an inaccurate response either because a truthful answer is likely to embarrass the respondent or because the respondent feels prompted to answer a certain way (from social desirability or acquiescence). The method of sampling is chosen to ensure the maximum return for the investment of time and money. The sampling frame is made as complete as possible to guard against bias in the results. Results are evaluated to find inconsistencies and contradictions. Similar surveys are examined to search for potential problems. In effect, well-done surveys involve a great deal of quality control.

The primary use of many surveys is to estimate the proportion of people in a population that fall into a certain category or categories. Even with good surveys, there are problems that occur during sampling and decisions made about the questionnaire that will perturb the estimate of this proportion. We will consider the problem of nonresponse and the effects of varying the number of possible responses on a questionnaire.

BACKGROUND INFORMATION

One of the most difficult problems in survey research is nonresponse. It is a mistake to assume that people who did not respond to a question are identical to those who did. Perhaps some people feel threatened by a question and therefore do not respond, while those not threatened by the question answered fairly neutrally. Perhaps the question excited some who responded in the extreme and bored others who did not respond. Perhaps there is some aspect of personality that a particular survey touches and divides the population into two diverse groups. Or perhaps the nonrespondents are just a cantankerous lot whose opinions we still need.

Another aspect of a questionnaire that will affect the response is the set of possible responses. One standard type of response is the Likert scale. A basic Likert item asks an individual to express a degree of agreement to or support of a statement on an ordered scale. For example, consider the statement "The use of marijuana for medicinal purposes should be legal." People might be asked if they agree, are neutral, or disagree with that statement. This is a 3-point Likert scale. But they could just as easily be asked on a 5-point Likert scale, i.e., if they strongly agree, agree, are neutral, disagree, or strongly disagree with that statement. Or if they strongly agree, agree, mildly agree, are neutral, mildly disagree, disagree, or strongly disagree with that statement. As another example of a Likert scale, many companies do product satisfaction surveys. Purchasers of the company's product are asked how satisfied they are with the product. They are instructed to indicate their degree of satisfaction by selecting a number between 1 and k, where a "1" would indicate complete dissatisfaction and "k" would indicate complete satisfaction. This is a k-point Likert scale. With the emphasis on customer satisfaction, many companies use this type of question to evaluate the quality of products or services. Often companies want to know what proportion of people check the "top box," i.e., what percent of their customers are completely satisfied and check the kth response. One interesting question asks, "What are the effects of changing the value of k?" That is, what happens to the proportion of people who check the top box as k is varied from 3 to 5 to 7, etc.?

QUESTIONS OF INTEREST

The general problem of nonresponse has been considered. Cochran [Cochran, 1977] discusses this problem in his sampling text. He divides the population into two strata, the first consisting of measurements from the respondents in a sample, the second being the nonrespondents. Of course, there are no measurements from stratum 2, the nonrespondents. Following the notation of Cochran, we let N_1 and N_2 be the number of members in the two strata. Then the proportions of the population in each stratum are $W_1 = N_1/N$ and $W_2 = N_2/N$, where $N = N_1 + N_2$. Suppose we let $\overline{Y}$ represent the mean of the population and $\overline{Y}_i, i = 1, 2$, the means of the two strata. Upon taking a simple random sample from the population, we have an estimated value for $\overline{Y}_1$, $\overline{y}_1$, but none for $\overline{Y}_2$. The resulting bias in the estimation of $\overline{Y}$ is

$$E(\overline{y}_1) - \overline{Y} = \overline{Y}_1 - \overline{Y} = \overline{Y}_1 - (W_1\overline{Y}_1 + W_2\overline{Y}_2) = W_2(\overline{Y}_1 - \overline{Y}_2).$$

We have no information on $\overline{Y}_2$. So for a continuous variable, we can place no reasonable bounds on the possible error. However the situation is better when dealing with categorical data. We want to estimate P, the proportion of the population that belongs in the top box. Let p_1 represent the proportion of successes in the observed sample, i.e., the proportion of successes among the respondents. If $W_1 = 1$, i.e., if there is no nonresponse, and if the sample was large enough to ignore the finite population correction factor, then approximate 95% confidence limits for P are given by

$$p_1 \pm 2\sqrt{\frac{p_1(1-p_1)}{n}}.$$

Because W_1 will not be 1, the bias introduced by the unobserved value of P_2 will perturb these limits. However, because we know that the value of any proportion must be between 0 and 1, we can use a conservative approach to create valid confidence limits by taking a "worst-case scenario" approach. For the lower limit (LL), the worst case would be if $P_2 = 0$, while for the upper limit (UL), the worst case is where $P_2 = 1$. Therefore, conservative limits are given by

$$LL = W_1\left(p_1 - 2\sqrt{\frac{p_1(1-p_1)}{n}}\right) + W_2(0),$$

$$UL = W_1\left(p_1 + 2\sqrt{\frac{p_1(1-p_1)}{n}}\right) + W_2(1).$$

These limits are very conservative and can lead to wide intervals. Cochran presents a table of LL and UL for $n = 1000$ in which the values of W_1 and p_1 are varied. Recognizing that W_2 is usually unknown (we will only have the nonresponse rate of a sample), we can again use a conservative approach. We calculate the LL by assuming that none of the n_2 nonrespondent observations belong in the top box and UL by assuming that all of the n_2 nonrespondent observations belong in the top box. If x represents the number of the n_1 respondents who check the top box, this gives

$$LL = \frac{x}{n_1 + n_2} - 2\sqrt{\frac{\left(\frac{x}{n_1 + n_2}\right)\left(1 - \frac{x}{n_1 + n_2}\right)}{n_1 + n_2}},$$

$$UL = \frac{x + n_2}{n_1 + n_2} + 2\sqrt{\frac{\left(\frac{x + n_2}{n_1 + n_2}\right)\left(1 - \frac{x + n_2}{n_1 + n_2}\right)}{n_1 + n_2}}.$$

These are the formulae used in the example on the bottom of p. 362 of Cochran.

The problem that Cochran presents is based on estimating a binomial proportion. This proportion, while unknown, is constant. This is not so with a Likert scale until the value of k is selected. If we are interested in the proportion of people who belong in the top box, this proportion depends on the choice of k in a k-point Likert scale. As the number of possible responses goes up, the proportion of people who are in the top box should go down. A person who is happy with his product and would willingly check 3 (satisfied) on a 3-point Likert scale might be reticent to check 5 (very satisfied) on a 5-point Likert scale, 7 (extremely satisfied) on a 7-point Likert scale, or 27 (very extremely infinitely magnificently bodaciously satisfied) on a 27-point Likert scale. Choosing the value of k actually changes the parameter being estimated. But many companies who sponsor surveys want to know the proportion in the top box, no matter the number of choices available. So the selection of k is crucial.

Combining the effects of nonresponse with this varying definition of the parameter of interest, an interesting question arises: With what values of k does the problem of nonresponse cause the greatest difficulties in estimation? This general question has interesting parts:

1. When are the widths of confidence intervals most affected?
2. How conservative are the confidence limits presented in Cochran?
3. When are the estimators most likely to be biased?
4. What are the effects of ignoring nonresponse?

DATA

To try to examine the combined effects of nonresponse and the choice of k in a k-point Likert scale, we suggest a computer simulation to generate data. Such simulations are often used in research to evaluate the statistical properties of a procedure. The simulation we suggest would generate random data that (our experience says) is typical for customer satisfaction. For most of these surveys, the distribution seems to have a positive mode with a negative skew [Peterson and Wilson, 1992]. A key point of the simulation is to use a probability model for the responses that would have this basic shape but would be flexible enough to be appropriate and in some sense comparable for different values of k. That is, we would like to select a probability model that would allow us to vary k yet still consider the samples to be from a population with the same basic shape. One choice would be to pattern the responses as samples from binomial populations. To ensure the correct shape to the distribution, we would require that the probability of success on an individual trial be greater than 0.5. (Note: Because we want to avoid confusion between this probability and the probability of membership in the top box, we will denote this binomial probability as π_R, while the probability of membership in the top box will be represented by P.) Then the various choices of k can be modeled by varying the value of n, where n represents $k - 1$.

For example, if we choose $n = 2$, we are modeling a 3-point Likert scale. The possible outcomes of the binomial sample are 0, 1, and 2. We will let a 0 represent 1 on the Likert scale, while 1 represents 2 and 2 represents 3. The extension to any values of n and k is obvious.

After we have selected the type of population to use to represent the respondents, we want to do the same for the population of nonrespondents. That is, we want to select a probability model, different from the population of respondents, to represent the population of opinions of the nonrespondents. The appropriately weighted combination of the two distributions would represent the distribution of the population. It is this combined population whose proportion of observations in the top box is the parameter of interest. (Note: We want the distribution of the nonrespondents to be different from that of the respondents because if they were identical, the only effect of nonresponse is a reduction in sample size.) Some reasonable choices for the simulation might be a second binomial with a parameter value π_{NR} different from π_R or a discrete uniform distribution.

ANALYSIS

Once the distributions are selected, we can vary two parameters. One is the value of k on the k-point Likert scale. The other is what Cochran denotes as W_2, the proportion of nonresponse. Cochran presents a table of 95% confidence limits for P for various values of P and nonresponse percents only up to 20%. We feel that this table presents too rosy a picture for the nonresponse rate, especially for but not limited to mail surveys, and feel it is more appropriate to consider nonresponse rates up to at least 50%.

For the analysis, we suggest a series of computer simulations. The first set would be directed by our choices for the distributions to use. The remaining simulations would be based on the reader's judgments on how to best further investigate the effects of nonresponse and selection of k. The first two simulations involve sampling from binomial distributions, the first with $\pi_R = 0.60$ and the second with $\pi_R = 0.75$. The remaining steps are identical.

1. Decide on the distribution of nonrespondents. We would suggest three reasonable possibilities: (1) binomial with parameter $\pi_{NR} = 1 - \pi_r$; (2) binomial with parameter $\pi_{NR} = 0.50$; (3) discrete uniform.

2. For $k = 3, 5, 7$ ($n = 2, 4, 6$), generate a sample of $1000 + s$ observations from the population of respondents and s observations from the population of nonrespondents. Do this for $s = 50, 100, 150, 200, 300$, and 500, corresponding to $W_2 = 5, 10, 15, 20, 30$, and 50%.

3. For each of these obtain the following:
    A. The true value of P, the proportion in the combined population that belongs in the top box.
    B. The conservative 95% confidence interval used at the bottom of p. 362 of Cochran.
    C. A 95% confidence interval using only the responses and thereby ignoring the problem of nonresponse, as is, unfortunately, often done in practice.
    D. A 95% confidence interval using both the responses and the data on the nonrespondents as if they were available. This combined confidence interval is the "right" confidence interval, i.e., the one we would calculate if the problem of nonresponse did not exist.

4. Summarize the results of the simulations in tabular form.

5. Examine the results in the table to get an initial impression of the effects of nonresponse and k and to suggest further simulations that would be beneficial.
6. Perform the additional simulations.
7. Repeat.

INSTRUCTIONS FOR PRESENTATION OF RESULTS

We will give an example of the process and a way that the results could be presented. For this example, we will present the results of a simulation where we used $\pi_R = 0.75$ and $n = k - 1 = 4$. We also let our population of nonrespondents have a binomial distribution, but with $X_{NR} = 0.25$. We used MINITAB to generate the samples and calculate the confidence intervals. For each value of W_2, we also calculated P, the proportion of the population in the top box. To do so, we calculate the proportion of respondents and nonrespondents that belong in the top box and take a weighted average. For example, for $W_2 = 10\%$, we obtained the following:

Respondents:

$$P(x=4) = P(\text{top box}) = \binom{n}{x} \pi_R^x (1 - \pi_R)^{n-x} = \binom{4}{4}(0.75)^4 (0.25)^0 = 0.3164 .$$

Nonrespondents:

$$P(x=4) = P(\text{top box}) = \binom{n}{x} \pi_{NR}^x (1 - \pi_{NR})^{n-x} = \binom{4}{4}(0.25)^4 (0.75)^0 = 0.0039 ,$$

$$P = .90(0.3164) + .10(.0039) = .285 .$$

The results are given in Table 1.

Table 1. *Results: $k = 5$, $\pi_R = 0.75$, $\pi_{NR} = 0.25$.*

W_2	P	Conservative	Respondents	Combined
5%	.301	.274	.289	.274
		.383	.349	.332
10%	.285	.253	.281	.254
		.412	.343	.310
15%	.270	.242	.286	.242
		.451	.350	.298
20%	.254	.221	.277	.221
		.479	.343	.275
30%	.223	.195	.281	.198
		.553	.351	.250
50%	.160	.119	.241	.122
		.671	.322	.167

Once this table and similar ones are obtained, the results of the tables should be compared, further simulations performed, and the results, with conclusions, presented in the form of a report.

REFERENCES

Cochran, William G. (1977), *Sampling Techniques*, Wiley, New York, pp. 359–362.
Peterson, Robert A. and Wilson, William R. (1992), *Measuring Customer Satisfaction: Fact and Artifact*, Journal of the Academy of Marketing Science, Vol. 20, No. 1, pp. 61–71.

BIOGRAPHIES

Dr. Smidt earned a B.S. in Mathematics from Manhattan College, an M.S. in Statistics from Rutgers University, and a Ph.D. in Statistics from the University of Wyoming. He has taught at the University of Florida, Rutgers University, Oregon State University, Vandenberg AFB, and the California Mens Colony. Dr. Smidt is interested in statistical consulting and applied statistics, including multivariate statistics, experimental design, regression, and survey sampling. He helped establish the university-wide statistical consulting service at Cal Poly and has extensive statistical consulting experience within the university. He has also worked with the Departments of Defense and Energy, Bechtel, Lawrence Livermore National Laboratory, Cogimet, and Lindamood-Bell. Dr. Smidt is Professor and Chair of the Statistics Department at Cal Poly in San Luis Obispo, CA.

Dr. Tortora joined Gallup as its Chief Methodologist in July 1995. Prior to joining Gallup, Dr. Tortora was the Associate Director for Statistical Design, Methodology and Standards at the U.S. Bureau of the Census. In this position, he was leading the design of the 2000 Census and was responsible for statistical and survey methodology for all Census Bureau programs, development of statistical and survey standards, and career management and training programs for mathematical statisticians and survey methodologists. Prior to assuming this position in 1992, he was chief of the bureau's Statistical Research Division. From 1975 until 1990, he served the United States Department of Agriculture in various positions until becoming the Director of the Research and Applications Division. Dr. Tortora also served as an Adjunct Professor of Applied Statistics at George Mason University in Fairfax, Virginia, as a lecturer in Advanced Survey Methods for USDA's Graduate School in Washington, DC. He is currently an Adjunct Professor at the University of Nebraska, Lincoln. He is an elected Fellow of the American Statistical Association.

CHAPTER 12

DESIGNING AN EXPERIMENT TO OBTAIN A TARGET VALUE IN THE CHEMICAL PROCESSES INDUSTRY

Michael C. Morrow, Thomas Kuczek, and Marcey L. Abate

The case history presents the issues encountered when designing an experiment. The emphasis is on the design, not the analysis. A proper design makes the statistical analysis straightforward. A lot of design issues are presented. An appropriate design option is chosen and evaluated for sensitivity given the constraints of the process. The data from the experiment is analyzed and summarized. Conclusions from the planning process and analysis are presented. It is hoped that students exposed to this case study will get a taste of what experimental design is truly about.

INTRODUCTION

The problem to be solved is to identify the critical variables involved in a chemical process and then to come up with an appropriate experimental design which will help put the process on target. The key issues are setting the objectives of the experiment and then choosing the design to achieve the objectives of the experiment. The emphasis in this case history is on the planning process, although the analysis of the chosen design is also presented.

BACKGROUND

A major goal in the production of plastic pellets at Eastman Chemical Company is to keep a property of the plastic pellets, in this case Response, as close to a target value as possible. It is critical that the response of the plastic pellets produced be close to the target value, for if it is not, the pellets cannot be used efficiently in the manufacturing processes of Eastman's customers. Eastman's customers use the pellets to produce sheeting, containers, refrigerator components, display cases, and so forth. If the response were to deviate from the target response value by a high enough margin, the result could be unsellable material or a substandard product if the pellets were used.

The plastic pellets are manufactured in a continuous process. The substrate material from which the pellets are formed is put in a charge bin and fed through to the first heater at the desired production rate. The material is then passed through a second heater and then into a production vessel. There the material is held under heat and pressure and mixed with catalysts for a while and then is transported to holding silos before being shipped to customers.

Because the process is required to produce plastic pellets with a response as close as possible to the target value of 30, Eastman engineers want to know how response changes when certain input factors vary. The engineers know that many inputs affect the response of the pellets which are produced. Some of these inputs can easily be controlled in the production process. These are generally referred to as control factors (or variables). There are other inputs which affect response which cannot be controlled in the production process. They are referred to as uncontrollable factors. The problem is how to find a way to set the control factors so that response remains at the target value of 30 even when the uncontrollable factors vary. The engineers also want to determine the effect of changing the factors on response. In order to accomplish this, an experiment needs to be designed to better understand how response depends on uncontrollable and controllable factors.

QUESTIONS OF INTEREST

There are many considerations in planning a designed experiment. For this case, the issues are summarized as follows:
- What are the objectives of the experiment? The objectives should be stated in engineering and statistical terms. The objectives must be clearly understood so that the experiment can be planned to obtain them.
- How many experimental runs will be needed at each setting? There are many considerations when determining the number of runs, one of the most important being the expense of each observation. Other issues to consider are listed below:
  - Is the process stable in terms of the important outputs? A stable process gives a predictable estimate of variability for use in the planning process. The results of a designed experiment are predictable if the process is stable during the experiment and remains stable after the experiment.
  - What is the variability of the important outputs? The standard deviation of the output measures the magnitude of the variability in the process. The effects of factors must be evaluated relative to the process variation.
  - What kind of model is needed (linear effects, interactions, quadratics)? The design must be adequate to fit the desired model.
  - What should the risks be set at (i.e., Type I (Alpha) and Type II (Beta) error probabilities)? The experiment should have a reasonably high probability of detecting important effects from an engineering standpoint, with a low probability of detecting an effect when it really does not exist.
  - At what factor levels should we collect data from the process? It is important to choose levels distinct enough to cause changes in the process that are statistically detectable, but the factor levels should be such that the process can still operate.

It takes a lot of work to plan an experiment correctly. However, it is the most important part of any experimental design. In most cases, if the experiment is planned correctly, the analysis will be trivial. Gerald Hahn put it very well in saying, "The world's best statistical analysis cannot rescue a poorly planned experimental program." Another relevant quote from a unknown source said, "A well-designed and executed experiment is usually more important than 'high-powered' statistical methods and will usually result in a trivial analysis with clear, unambiguous conclusions." More detail now follows on the points mentioned above.

Objectives

In order to assure that the experiment is properly designed to meet all of the objectives, they must be clearly stated prior to the actual implementation of the experiment. As stated earlier, the engineering objectives are as follows: (1) Develop a control strategy for response; (2) Determine the effect of changing important inputs on response. The statistical objective is to develop a model to estimate response as a function of the input factors. The model will help develop the control strategy and evaluate the effect of changing the important inputs on response.

Sample Size

"How many samples do I need to run?" is probably the most frequently asked question of a statistician. From a statistical viewpoint, there are many considerations when determining a proper sample size. Some of the most important ones are

- Is the process stable?
- What is the variability of the response?
- What is the minimum effect the experiment needs to detect?
- What are the alpha and beta risks?
- What kind of model is needed?

The sample size is estimated using the information gained from answering these questions. However, there are also important considerations such as the cost associated with each sample. For the plastic pellets, the process is manufacturing costly material at every combination of input factors (experimental conditions).

In addition, there are often engineering constraints. For example, how long does it take for the process to line out? That is, after making a change to the input factors, how long does it take for the process to give representative output values (in this case, response)? For the plastic pellet manufacturing process, the engineers estimated the line-out time to be 48 hours. After the process lines out, the process can stay at the experimental conditions for 24 hours. During this 24 hours, one data point can be collected every four hours. Thus, each combination of input factors will be held for 72 hours and will provide six data points. The engineers decided they could run a maximum of 16 sets of experimental conditions, resulting in an experiment which took 48 days. Given these constraints, the sample size question becomes one of determining experimental sensitivity. That is, given that six data points can be collected at each of 16 sets of experimental conditions, what is the minimum effect the experiment can detect?

Stability

It is critical that the process be stable in order to obtain a predictable estimate of variability and predictable results from a designed experiment. At each set of conditions, six individual response measurements will be averaged and used to develop a model. Therefore, we must determine if the process is stable with respect to the averages over time. Figure 1 displays an individuals chart on the averages over time. This chart was chosen over an X-bar chart since it will include day-to-day variation in the stability evaluation. Because this variation will be included in the root mean square error from the model, it is appropriate to include it in the stability evaluation and sample size calculation. There is also some question as to the independence of the individual measurements within a day. Time series analysis indicated that the individual measurements were not independent. Therefore, it is not appropriate to construct control limits for an X-Bar chart with nonindependent data in the subgroups. The individuals and moving range charts indicate instabilities in the process. If the root causes for the instabilities are identified and removed, the process will operate with less variability. This is accomplished through ongoing statistical process control (SPC) efforts. The engineers decided that the degree of instability was not bad enough to stop the experiment.

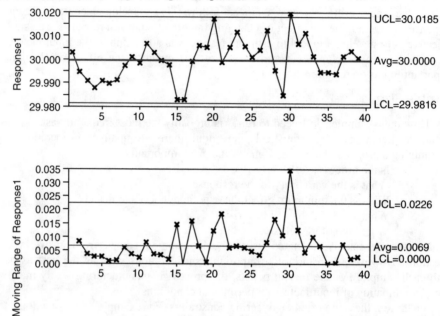

Fig. 1. *Individuals control chart on response averages.*

Process Variability

The process variability is an important input in determining the number of replicates of each experimental setting or in determining experimental sensitivity. The standard deviation is a measure of the process variability. Because the root causes of the instabilities are still part of the manufacturing process, all of the data will be used to calculate the standard deviation of the daily response averages. Figure 2 displays a histogram and descriptive statistics of the daily response averages. The standard deviation of the daily averages is 0.00864.

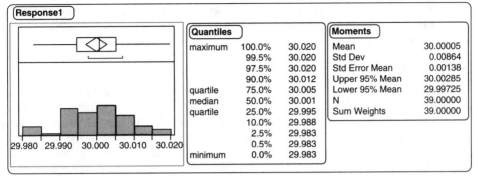

Response1					
Quantiles			**Moments**		
maximum	100.0%	30.020	Mean		30.00005
	99.5%	30.020	Std Dev		0.00864
	97.5%	30.020	Std Error Mean		0.00138
	90.0%	30.012	Upper 95% Mean		30.00285
quartile	75.0%	30.005	Lower 95% Mean		29.99725
median	50.0%	30.001	N		39.00000
quartile	25.0%	29.995	Sum Weights		39.00000
	10.0%	29.988			
	2.5%	29.983			
	0.5%	29.983			
minimum	0.0%	29.983			

Fig. 2. *Histogram and sample statistics from the average response data.*

Model Type

The type of model the experimenters want to estimate will help determine the experimental design. The model type will also be used as an input to determine the sensitivity of the experiment. Using a computer program based on the article in [Abate, 1995], the minimum detectable effect of the experiment can be calculated in multiples of the standard deviation. The engineers decided it was important to estimate interactions. They also wanted to estimate nonlinear relationships.

Risks

The experimenters must choose the risk levels in an experiment. The majority of experimenters have a difficult time choosing Type I (Alpha) and Type II (Beta) risk levels. Alpha is the risk of concluding that an effect of a factor on response exists when it truly does not. Beta is the risk of concluding that an effect of a factor on response does not exist when it truly does. Experimenters will often set alpha = 0.05 and beta = 0.10 because these choices are given as a rule of thumb in many textbooks. However, more informed risk decisions may be made by considering such things as

- At what stage is the experimental process?
  - Are a large number of factors being screened for importance?
  - Are the key factors being optimized?
- What are the implications to the customer if a wrong decision is made based on the experimental data?

Sensitivity

The experiment should be able to detect a 0.0169 change in response across the range of Z_1(uncontrollable factor), X_1, and X_2 (controllable factors). This calculation was done using the method described in [Abate, 1995] and the information above. The engineers decided the sensitivity of the experiment was acceptable. If it were not acceptable, they would have several options:

- They could replicate some or all of the experimental conditions.
- They could leave the process at the experimental conditions longer and collect more four-hour samples to go into the average.
- They could work on reducing the variation in the process. In this case, nested experimental designs are useful to determine the major sources of variation. One can use this information to determine where further work is needed to reduce the variation.

- They could also decide that the sensitivity of the experiment is unacceptable and run the experiment anyway. In this case they are hoping the experiment will produce shifts in response greater than the detectable effect.

Factor Levels

Factor levels must be determined by the people knowledgeable of the process, typically the engineers and operators. However, statisticians can offer some considerations such as the following: (a) The range of the factor levels should be large enough to induce large changes in response relative to the variation in the process; (b) However, the ranges must be chosen such that the process can be run at the experimental settings. One issue the engineers brought out at this stage was the problem of running at extreme conditions, which a typical factorial type of design requires. Catching a plant on fire does nobody any good! Most engineers cannot make "off class" or unsellable material during an experiment.

DATA

A three-factor Box–Behnken experiment was designed and run on the process. A Box–Behnken was chosen because the client wanted to evaluate curvature terms and felt that running the process at the extreme vertices of a factorial design would not be possible. Three replicates were added at the center point conditions to give an estimate of pure error and to track the process stability during the experiment. The center points will be spaced evenly throughout the experimental runs. The rest of the combinations were selected randomly. However, the engineers had to change the run order slightly for a variety of practical reasons. The run order column gives the planned order for the experimental conditions. The actual order of the conditions is how the data is ordered in the file. In addition, only three data points were obtained for run 12 because the response was getting dangerously high, which forced leaving the conditions before obtaining all six data points.

The data is contained in the Case12.txt file (see Tables 1 and 2). At each of the 16 design conditions, product was manufactured for one day. Every four hours, a sample of material was taken and sent to the lab for analysis. The response and the experimental factors have been coded in order to protect proprietary information. Although this data set is real, Eastman's interest must be protected. Therefore, the data can only be given out in coded form. No more information can be given on the details of the data.

Table 1. *Variable descriptions.*

Var Name	Description
Response	Important quality characteristics for Eastman's customer
Z_1	Uncontrollable process variable
X_1	Controllable process variable 1
X_2	Controllable process variable 2

Table 2. *First three cases of Case12.txt.*

Run_Order	Response	Z_1	X_1	X_2
1	30.010	0	0	0
1	29.995	0	0	0
1	30.009	0	0	0

ANALYSIS

A least squares fit of the data was made. The model contained linear, quadratic, and interaction (or cross product) terms. The refined model with terms that had a p-value of 0.10 or less is listed below.

$$R = 30.002 + 0.018X_1 + 0.019X_2 - 0.022Z_1 - 0.009X_1*Z_1 + 0.01Z_1^2 - 0.008X_2^2 ,$$

where R is the predicted response for the fitted model (see [Bratcher, Moran, and Zimmer, 1970]).

From the fitted model, an engineer may gain insight into the relationships among these factors in addition to an empirical ability to predict response as a function of these variables. The plots in Figures 3 and 4 display the relationships between response and the factors in the experiment. Figure 3 shows how the quadratic relationship between the response and Z_1 changes with X_1. When X_1 is high the rate of decline in response is significantly greater in the lower range of Z_1 than when X_1 is low. Or, the effect of changing X_1 from low to high is significantly greater when Z_1 is low than when Z_1 is high. This information is very useful to the engineers for process control. Typically engineers understand how one variable generally effects a response. But, it is usually new information to them when two variables interact. Figure 4 shows how response increases as X_2 increases. It shows how response begins to level out as X_2 reaches the high levels.

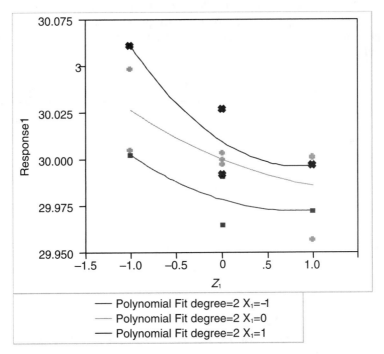

Fig. 3. *Response vs. Z_1 by X_1 (across all levels of X_2).*

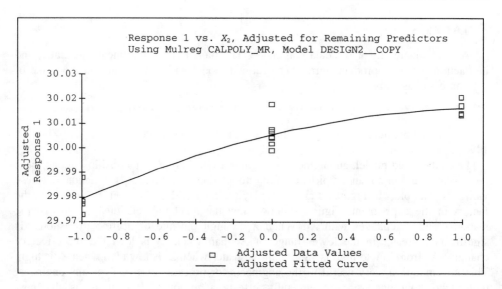

Fig. 4. *Response vs. X_2.*

A revealing outcome of the analysis comes when one considers the practical implications of the model at different levels of Z_1. One can scan the surfaces for specific settings of X_1 and X_2 to keep the process on target (response = 30) as Z_1 varies. Figures 5, 6, and 7 display the contour plots of response versus X_1 and X_2 as Z_1 is fixed at the low, medium, and high levels.

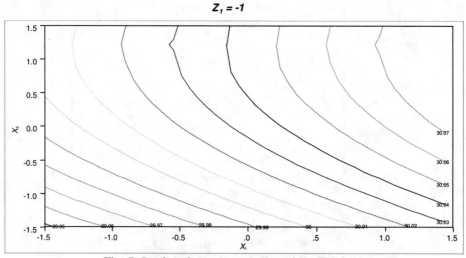

Fig. 5. *Predicted response vs. X_1 and X_2 (Z_1 - low).*

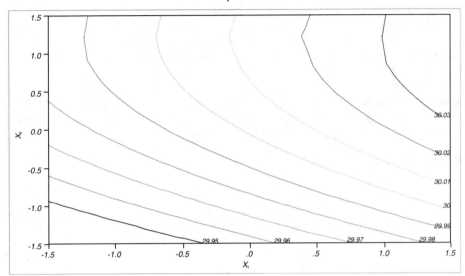

Fig. 6. *Predicted response vs. X_1 and X_2 (Z_1 - middle).*

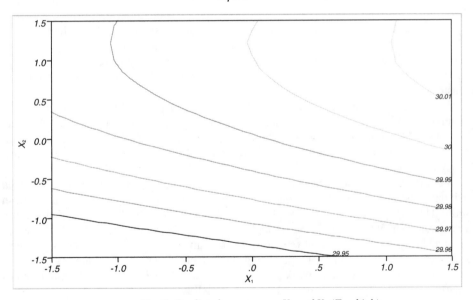

Fig. 7. *Predicted response vs. X_1 and X_2 (Z_1 - high).*

CONCLUSIONS

A three-factor Box–Behnken experiment was chosen. The Box–Behnken was chosen for three reasons: (1) A quadratic model was needed; (2) Operations was nervous about operating at the extreme conditions of a factorial design; (3) Minimization of the number of conditions run on the process was desired. The last reason is always a concern for practitioners. A statistician needs to help them understand how sensitive the experiment will be given the sample size constraints, variation, and objectives.

The goal is to put the process on a target value of 30. To do this, we ran a designed experiment to see how three factors of interest affect response. Our empirical model does this and allows us to estimate response for any setting of variables within the range studied. The engineers will be able to maintain the target response while Z_1 varies by changing the settings of X_1 and X_2. This is extremely useful information to the engineers. They will be able to operate the process on target, and hence with less variability as an uncontrollable factor (Z_1) varies in time. The information gained from this experiment will give Eastman Chemical Company a competitive advantage in the manufacturing of plastic pellets.

REFERENCES

Abate, Marcey L. (1995), *The Use of Historical Data in Statistical Selection and Robust Product Design*, Ph.D. Thesis, Department of Statistics, Purdue University, West Lafayette, IN.

Bratcher, T.L., Moran, M.A., and Zimmer, W.J. (1970), *Tables of Sample Sizes in the Analysis of Variance*, Journal of Quality Technology, Vol. 2, No. 3, July, pp. 156–164.

Mason, Robert L., Gunst, R.F., and Hess, H.L. (1989), *Statistical Design and Analysis of Experiments*, New York: John Wiley and Sons, Inc., pp. 220–228.

Neter, John, Wasserman, W., and Kutner, M. H. (1990), *Applied Linear Statistical Models*, Richard D. Irwin, Inc., p. 223, Homewood, IL.

BIOGRAPHIES

Tom Kuczek received his Ph.D. from Purdue University in 1980, then returned there where he is currently Professor of Statistics and Associate Head of the Statistical Consulting Service. Tom teaches Experimental Design, Response Surface Methodology, and Statistical Process Control. Tom's research interests include Response Surface alternatives to Taguchi Methodology, modeling drug response and resistance of cancer cells to Chemotherapeutic agents, and applications of Statistics to problems in Civil and Industrial Engineering.

Mike Morrow received his Master of Science in Applied Statistics from Purdue University in 1991. He currently works as a Statistical Consultant for Eastman Chemical Company. Mike primarily consults with clients in the applications of Experimental Design techniques in Research, Development, Analytical Methods, and Manufacturing at Eastman Chemical Company.

Marcey Abate is a member of the Statistics and Human Factors Department at Sandia National Laboratories in Albuquerque, New Mexico. She has a Ph.D. in Statistics from Purdue University, is an American Society for Quality Control (ASQC) Certified Quality Engineer, and is active in the ASQC Statistics Division. Her interests include the design and analysis of experiments, the reliability of measuring instruments, statistical graphics, experiments, the reliability of measuring instruments, statistical graphics, and risk assessment. Marcey Abate's work was partially supported under the United States Department of Energy under contract DE-AC04-94AL85000.

Analysis and designs done using RS/Discover software. RS/Discover is a product of BBN Inc.

Analysis and designs done using JMP® software. JMP® is a product of SAS Institute.

CHAPTER 13

Investigating Flight Response of Pacific Brant to Helicopters at Izembek Lagoon, Alaska by Using Logistic Regression

Wallace P. Erickson, Todd G. Nick, and David H. Ward

Izembek Lagoon, an estuary in Alaska, is a very important staging area for Pacific brant, a small migratory goose. Each fall, nearly the entire Pacific Flyway population of 130,000 brant flies to Izembek Lagoon and feeds on eelgrass to accumulate fat reserves for nonstop transoceanic migration to wintering areas as distant as Mexico. In the past 10 years, offshore oil drilling activities in this area have increased and, as a result, the air traffic in and out of the nearby Cold Bay airport has also increased. There has been a concern that this increased air traffic could affect the brant by disturbing them from their feeding and resting activities, which in turn could result in reduced energy intake and buildup. This may increase the mortality rates during their migratory journey. Because of these concerns, a study was conducted to investigate the flight response of brant to overflights of large helicopters. Response was measured on flocks during experimental overflights of large helicopters flown at varying altitudes and lateral (perpendicular) distances from the flocks. Logistic regression models were developed for predicting probability of flight response as a function of these distance variables. Results of this study may be used in the development of new FAA guidelines for aircraft near Izembek Lagoon.

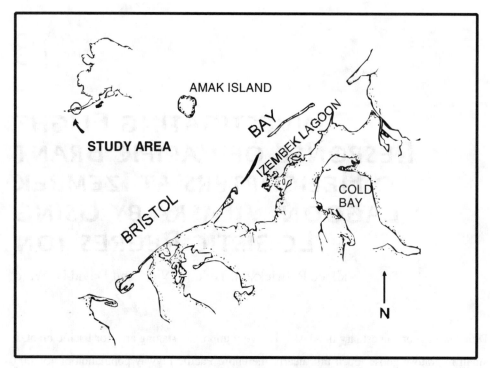

Fig. 1. *Location of study area (Izembek Lagoon).*

INTRODUCTION

During the fall season, Izembek Lagoon, Alaska (see Fig. 1) and adjacent estuaries support greater than 90% of the Pacific black brant (Branta bernicla nigricans) population. Izembek Lagoon was designated as a wetland of international importance in 1985 because it supports large numbers of geese and other waterbirds. Aircraft and other human activities may increase in and near Izembek Lagoon if petroleum industry facilities and a U.S. Coast Guard search and rescue station are established in Cold Bay. Brant and other geese are sensitive to aircraft and other human disturbance during fall migration. Human disturbance can disrupt feeding activity of geese, displace birds from feeding areas, and potentially affect energy reserves that are important for migration and over-winter survival of waterfowl. The current FAA minimum altitude standard for flying over Izembek Lagoon is 2000 feet.

To minimize potential impacts that may result from increased aircraft disturbance at Izembek Lagoon, an understanding of factors that influence the response of brant is needed. [Ward, Stehn, and Derksen, 1994] found that brant flew longer when disturbed by helicopters than by any other aircraft, but response varied by aircraft altitude and distance to the birds during fall-staging at Izembek Lagoon. Other studies (e.g., [Davis and Wiseley, 1974], [Owens 1977]) have also identified aircraft type, altitude, and distance from the birds as important factors influencing the response of waterfowl, yet few researchers have made a detailed examination of the relationship of aircraft altitude and distance to the birds and the response of brant.

The United States Fish and Wildlife Service is interested in developing models of the behavioral response of brant to aircraft as a function of altitude and lateral distance. The purpose of this study is to estimate the flight response of brant as a function of altitude and lateral distance using logistic regression techniques and provide recommendations for changes to the current FAA regulations if these appear inadequate for minimizing the effects to brant.

BACKGROUND

In this study, overflights of two large helicopters with similar noise ranges, the Bell 205 and a Sikorsky HH-3F Coast Guard helicopter, were flown following established routes and altitudes. These two aircraft are commonly used in the area and would be expected to have the highest disturbance effects on brant. Flight-lines were aligned perpendicular or parallel to the shoreline of Izembek Lagoon to simulate local flight patterns. Data presented are for these experimental aircraft overflights (hereafter called overflights) at Izembek Lagoon and were used to assess the effects of altitude of the aircraft and lateral (perpendicular) distance from the birds on the behavioral response of brant. Lateral distance was determined from maps, which included the flight path of the aircraft and the location of the center of each flock. The behavioral response of brant flocks to overflights was recorded from blinds at various locations along the shoreline of the lagoon. Flock size was determined by visual estimation from the blinds. A flock was defined as a spatially distinct group of birds. In some cases flock members were dispersed over a 1 km area, and an arbitrary subdivision of the flock was selected for observation.

QUESTIONS OF INTEREST

Investigate the effects of lateral distance and altitude of the helicopters on the flight response of brant using both graphical and statistical techniques. Develop a logistic regression model for flight response of brant as a function of the lateral distance and altitude and the interaction of the two. Investigate the fit of the models using appropriate logistic regression diagnostics.

DATA

Data collected on each brant flock during the experimental overflights include altitude of the helicopter during the overflight (ALTITUDE), lateral distance from the aircraft to the center of the flock (LATERAL), and flight response (FLIGHT). Descriptions, units, and ranges for these variables are provided in Table 1.

Table 1. *Variable names, descriptions, units, and ranges.*

Variable	Description	Units	Ranges Low	High
LATERAL	Lateral distance	100 meters (m)	0	80.47
ALTITUDE	Altitude of aircraft	100 meters (m)	0.91	12.19
FLIGHT	Flight response		0	1

Each overflight followed established routes and altitudes. Flight-lines were aligned perpendicular or parallel to the shoreline of Izembek Lagoon to simulate local flight patterns. Airspeed was maintained at normal cruising speed (150–240 kilometers/hour (km/hr)) in level flight. The altitude for an overflight (ALTITUDE) was randomly assigned (within levels tested) and was recorded at 9 discrete levels (0.91, 1.52, 3.05, 4.57, 6.10, 6.71, 7.62, 9.14, 12.19). Lateral (perpendicular) distance (LATERAL) between the aircraft and flock was determined from the study area maps to the nearest 0.16 km. Pilots of the aircraft radioed the observers with the altitude of the helicopter prior to initiating the next overflight. See Fig. 2 for an illustration of these measurements.

The behavioral response of goose flocks to overflights (FLIGHT) was measured from blinds at 8 study sites (see [Ward, Stehn, and Derksen, 1994]) and various temporary locations along the shoreline of the lagoon. A flock was defined as a spatially distinct group of birds; median flock size was 700 birds (range = 10 –30,000 birds). In some cases, flock members were dispersed over a 1 km area, and an arbitrary subdivision of the flock was selected for observation. The behavioral response was estimated as the percentage of the flock exhibiting flight response and was classified into two categories with 0 representing flocks exhibiting less than 10% response and 1 otherwise. Only two classes were used, because the majority of responses were either 0% or 100%. A flight response is defined as the brant individual taking flight during the overflight of the aircraft.

Table 2 contains the first 10 records of the data file Case13.txt, while Appendix A contains all 464 records of response by flocks of brant.

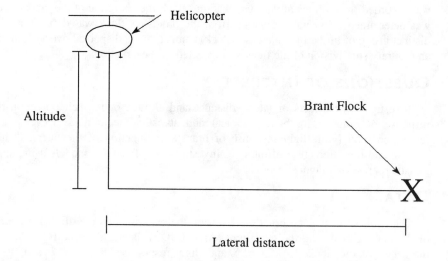

Fig. 2. *Illustration of the altitude and lateral distance measurements.*

Table 2. *First ten records of the brant response data file.*

FLOCK ID	ALTITUDE	LATERAL DISTANCE	FLIGHT RESPONSE
1	0.91	4.99	1
2	0.91	8.21	1
3	0.91	3.38	1
4	9.14	21.08	0
5	1.52	6.60	1
6	0.91	3.38	1
7	3.05	0.16	1
8	6.1	3.38	1
9	3.05	6.60	1
10	12.19	6.60	1

ANALYSIS

Phase 1. Describing data

Investigate the univariate distributions of altitude (ALTITUDE) and lateral distance (LATERAL) with the use of histograms and descriptive statistics. What is the shape of the distribution for LATERAL? Display frequency tables (frequencies and percentages) for the ALTITUDE and FLIGHT variables.

Visually compare ALTITUDE for the two categories (Yes/No) of FLIGHT with the use of side-by-side boxplots. Do the same for the LATERAL variable. What can you tell about your data from these boxplots? Does there appear to be a relationship between ALTITUDE and FLIGHT? How about LATERAL and FLIGHT? Summarize your results as if you were writing a report for a biologist.

Show a scatter plot of ALTITUDE and LATERAL. What is the scatter plot depicting about the design of the experiment? Is it a balanced design? Is there anything interesting about the altitude values greater than 12?

Phase 2. Odds Ratios and Simple Logistic Regression Analysis

This phase of analysis is designed to explore the relationships between the predictor and response variables separately. Models were fitted with ALTITUDE as the predictor, then with LATERAL as the predictor. This will demonstrate the relationships between the predictor variables and response, ignoring the effect of the other predictor.

Reduce the variable ALTITUDE to the three categories of <3, 3–6, and >6 and call the newly created variable ALT3CAT. Compute descriptive statistics (e.g., proportions) for ALT3CAT by flight response by using contingency tables (PROC FREQ in SAS).

When the response variable is a proportion (i.e., proportion of brant flocks exhibiting flight response), say p, it is often useful to apply the logit transformation: logit(p) = log($p/(1-p)$). This transformation is called the log odds ratio ($p/(1-p)$ are the odds of the brant flying). Two categories can be compared by taking the ratio of the odds ratios or log odds ratios. The resultant ratio of odds ratios yield the odds of a flock exhibiting flight behavior in the numerator category relative to the denominator category.

Is there an association between ALT3CAT and FLIGHT ignoring the effect of LATERAL? Calculate and discuss the odds and log odds for each category of ALT3CAT. Calculate and discuss the odds ratio for comparing the two altitude groups >6 and <3. Finally, fit a logistic model with ALTITUDE as the independent variable and FLIGHT as the dependent binary variable. Compute the estimates, standard errors, Wald chi-square statistics, and p-values for the intercept and ALTITUDE variable. Interpret the odds ratios for ALTITUDE from the logistic regression analysis.

Is there an association between LATERAL and FLIGHT ignoring the effect of ALTITUDE? To compute odds ratios only, collapse the LATERAL variable into four categories, <10, 10–19, 20–29, and >30 (LAT4CAT) and produce a contingency table for LAT4CAT by flight response. Calculate the odds and log odds for each category. Compute the odds ratio for the two groups <10 and >30. Does there appear to be an association between the LAT4CAT and FLIGHT? Using the continuous variable LATERAL, fit a logistic model with LATERAL as the independent variable and FLIGHT as the dependent variable. Is there as association between LATERAL and FLIGHT? Interpret the odds ratios for LATERAL from the logistic regression analysis.

Phase 3. Multiple Logistic Regression Analysis

Run a logistic model with ALTITUDE and LATERAL as independent variables and the flight response variable as the dependent variable. Are the coefficients significant? Interpret the odds ratios for both variables. Plot the predicted probability of flight response against LATERAL for each of three altitudes (2, 4, 6). With ALTITUDE fixed at 6.1 units (610 m or ~2000 ft), what value of LATERAL yields a predicted probability of flight response of 25%?

Consider the possible interaction between ALTITUDE and LATERAL. Plot the proportion of flocks responding to an overflight versus LATERAL (use the four groups defined above) separately for each of the three ALTITUDE categories defined previously (variable ALT3CAT). From this plot, does there appear to be an interaction between ALTITUDE and LATERAL? Include the interaction effect in the logistic model with ALTITUDE and LATERAL. What is the p-value for the interaction effect? Plot the predicted probability of flight response against LATERAL for each of the three altitudes (2, 4, 6) using this model.

With ALTITUDE fixed at 6.1 units (610 m or ~2000 ft), what value of LATERAL yields a predicted probability of flight response to helicopters of 25%? The current minimum altitude for all aircraft flying over the lagoon is 2000 feet (6.1 100 meter (m) units) when flying over the lagoon. Does this appear adequate if it is important to minimize the flight response of aircraft? Is there adequate information from this study to determine a minimum altitude requirement that yields a 25% probability of flight response by flocks when a large helicopter is flying directly over the lagoon? If not, and if studies were to be conducted to determine this altitude, what would you recommend in terms of study design? What would be your recommendation for setting minimum altitude or distance from the lagoon for these large aircraft if it was important to have a 25% or less chance of disturbing the brant flocks into flight?

Phase 4. Diagnostics

In linear regression, diagnostics using the residuals (i.e., difference between the fitted and the observed value) are investigated to identify outliers and data values having a large influence on the fit of the model. Several statistics have been developed for identifying outliers and influential observations in logistic regression (see [Pregibon, 1981] and [Hosmer and Lemeshow, 1989]) and can be calculated in most statistical packages for logistic regression. One such statistic, called the deviance residual, is commonly used. Which cases have large deviance residuals (define large as greater than 4)? What possible biological factors could be used to explain the ill-fitting observations? The sum of deviance residuals can be thought of as an overall fit for the model. Identify highly influential cases by calculating the change in the overall deviance statistic due to deleting that case.

INSTRUCTIONS FOR PRESENTATION OF RESULTS

A detailed written report that can be read and understood by a wide array of people, including statisticians, managers and staff, biologists, and environmental groups, should be submitted. A presentation that summarizes your analysis, results, and recommendations should be prepared. Assume your presentation has this same intended audience.

REFERENCES AND NOTES

Davis, R.A. and A.N. Wiseley (1974), *Normal behavior of snow geese on the Yukon-Alaska North slope and the effects of aircraft-induced disturbance on this behavior*, September,

1973; in W.W.H. Gunn, W.J. Richardson, R.E. Schweinsburg, and T.D. Wright, eds., *Studies of Snow Geese and Waterfowl in the Northwest Territories, Yukon Territory and Alaska,* 1973, pp. 1–85; Arctic Gas Biol. Ser., 27, 85 pp.

Hosmer, D.W., Jr. and S. Lemeshow (1989), *Applied Logistic Regression,* New York: John Wiley & Sons, Inc.

Owens, N.W. (1977), *Responses of wintering brant geese to human disturbance,* Wildfowl 28:5–14.

Pregibon, D. (1984), *Logistic regression diagnostics,* Annals of Statistics, 9, 705–724.

Ward, D.H., R. A. Stehn, and D.V. Derksen (1994), *Response of staging brant at the Izembek Lagoon, Alaska,* Wildl. Soc. Bull., 22: 220–228.

BIOGRAPHIES

Wallace P. Erickson is a biometrician with WEST Inc. He received his M.S. degree in statistics from the University of Wyoming in 1991. His primary interests are in the design and analysis of ecological studies, with primary emphasis on ecological risk assessment. He was a principal biometrician responsible for the design, conduct, and analysis of studies related to the Exxon Valdez oilspill in Prince William Sound, Alaska. He also has extensive experience in designing population abundance surveys and habitat selection studies of wildlife using methodologies such as line transect sampling, capture/recapture methods, and resource selection techniques.

David Ward is a research biologist with the US Geological Survey's Biological Resource Division in Anchorage. He received his early training at Hartwick College and later at the University of Oregon, where his master's research was on the wintering food habits of black brant. Since 1985, David has conducted fieldwork on arctic nesting geese in Alaska, Russia, and Mexico. His research has focused on population dynamics and migration ecology of waterfowl.

Dr. Nick is the biostatistician for the School of Health Related Professions at the University of Mississippi Medical Center in Jackson. He is in his first year at the associate professor level in the Department of Health Sciences, where he is a teacher and the statistical consultant for the school. He participates as a co-investigator in NIH and NCI grants with researchers from the medical center. He is a manuscript reviewer for many journals, including the Journal of Statistical Education, and was on the editorial board of the American Journal of Occupational Therapy. He was the 1997 Program Chair for the Section of Teaching of Statistics in the Health Sciences. He was President of the Mississippi Chapter of the American Statistical Association in 1995. Dr. Nick was also President of the Faculty Senate in 1997 at his University and was a Division Chairman for the Mississippi Academy of Sciences in 1996.

CHAPTER 14

ESTIMATING THE BIOMASS OF FORAGE FISHES IN ALASKA'S PRINCE WILLIAM SOUND FOLLOWING THE *EXXON VALDEZ* OIL SPILL

Winson Taam, Lyman McDonald, Kenneth Coyle, and Lew Halderson

The Alaska Predator Ecosystem Experiment (APEX) is a research project to determine why some species of seabirds whose populations were reduced by the *Exxon Valdez* oil spill in Prince William Sound, Alaska are not recovering. An acoustic survey was performed in the Sound to estimate the abundance and distribution of forage fishes and seabirds in the region. APEX involves a number of aspects, including estimation of seabird population sizes, food abundance, and state of the ocean. The sampling design was conducted with designated straight line paths transecting in three regions of the sound in July, 1995. These three regions were chosen to represent three levels of impact by the *Exxon Valdez* accident. The data consist of acoustic sonar signals collected on each transect using surface sensors, observer sightings of birds, net sampling of fishes, and water and weather conditions. This case study provides analysis of a segment of this study; namely, estimating the biomass of one species of forage fish with spatially correlated data. Other components of the project will evaluate the forage fish data collected in concert with seabird reproduction data over three years, 1995–1997, in an attempt to determine if food is limiting recovery of the piscivorous (fish-eating) seabirds.

INTRODUCTION

In a field study, many issues related to planning, execution, and analysis are crucial to the success of a project. Although the focus of this case study is placed on estimation of biomass, many interesting challenges can be found in the questions of interest and the comments and suggestions sections. In short, this case study provides analysis and estimation of the biomass of specific species of forage fish given that these measurements are likely to be correlated in space. Moreover, the location and concentration of these fish are reported as helpful information for scientists to monitor the ecological changes in this region.

BACKGROUND INFORMATION

Seabirds suffered substantial mortality from the *Exxon Valdez* oil spill in 1989, and some piscivorous species have not yet recovered. Low reproduction of piscivorous seabirds may be linked to food shortages related to oil spill injury. Since small forage fish are a major component of the diets of these seabirds in Prince William Sound, Alaska, the causes of seabird population declines cannot be evaluated without concurrent estimates of forage fish populations in foraging regions in proximity to the nest sites during the reproductive season. Three study regions (Figure 1) were selected to represent three levels of impact by the *Exxon Valdez* accident and were subsampled to provide estimates of forage fish distribution, species composition, size, and abundance using quantitative acoustic equipment from a research vessel. A second research vessel followed and captured fish from a sample of detected schools to provide estimates of the size and species composition of fish. The surveys were first initiated in 1994 in a pilot study to evaluate gear and sampling designs to be used in subsequent years. The seabird reproductive season of 1995 was the first full implementation of the sample survey; plans are in place to continue the surveys for two more reproductive seasons through the summer of 1997.

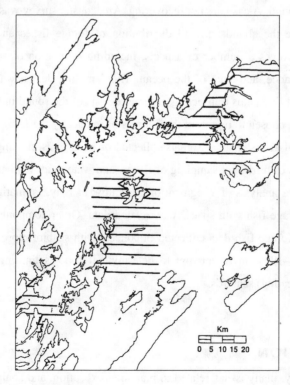

Fig. 1.

Each of the 3 study regions was subsampled twice during the 1995 seabird reproductive season using a systematic sample of parallel east–west transects. The data analyzed in this case study are from the first coverage of the transects in the middle study area (Figure 1), 1995. Two sets of similar data separated by about 2 weeks were collected from each of the 3 study areas. Figure 2 displays the 1995 systematic subsample of 12 survey lines (excluding the zigzag lines in Figure 1) which were traversed by the research vessels. The lines were selected across the central study region with a random starting point for the first line. The remaining 11 lines were then located parallel to the first line and at equal intervals of about 3.7 kilometers (2.3 miles) apart. Note that wind and waves make it impossible for a ship to exactly follow a line and the realized survey tracks were not exactly straight. Also, there are 6404 numbers in the data set for biomass of a certain age forage fish, 12 lines in the systematic set of lines, and 1 random placement of the systematic set. From these comments, one can begin to see that these data are not exactly like the standard "simple random sample" with "independently and identically distributed" random values.

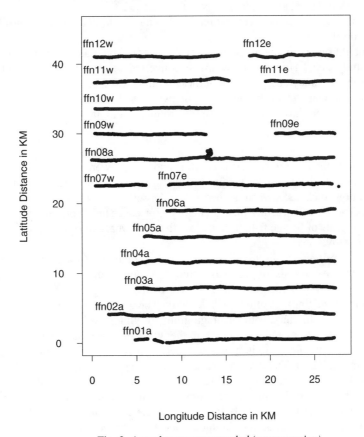

Fig. 2. *Actual transects sampled (center region).*

Our collaboration provides assistance with estimation of the biomass of specific species of forage fish given the fact that these data are likely correlated in space. Data from points close together in space are likely more similar than data from points which are separated by large distances. Other components of the project will evaluate the forage fish data and data of seabird reproduction over three years, 1995–1997, in an attempt to determine if food is limiting recovery of the piscivorous seabirds.

QUESTIONS OF INTEREST

One of the objectives of APEX is to assess the amount of forage fish available for the region's seabirds. Developing a procedure to graphically illustrate the distribution of a specific species of forage fish in one region and to estimate the total biomass of the species in subregions is the objective of our case study. This objective may seem simple but there are a number of questions associated with it.

1. What is the sampling procedure? What is the sample size? There are 6404 data records in the set, but is this the proper sample size? Is the sample size 12, the number of designated transects?
2. Some data values are from adjacent sites and others are separated by several kilometers. How does this affect the analysis? About 54% of the values for biomass are less than 0.001 g/m^2. How does this influence the estimation?
3. What are the underlying assumptions for estimation of mean biomass and total biomass? Are the data values independent?
4. What is the difference between estimating biomass with and without consideration of spatial correlation?
5. How should one compute an estimate of the average biomass per square meter in the entire study area? How is this estimate converted to an estimate of total biomass? What is the accuracy and precision of these biomass estimates?
6. How are estimates from two locations or two time frames compared to determine if important changes have taken place in the amount of biomass present?

DATA

In the study, three regions, north, central, and south, were selected from the Sound (Figure 1). The data include sonar measurements of biomass along the designated transects, fish information such as size and species, an observation log of the birds' foraging behavior, and ocean and weather measurements. Our case study concentrates on the sonar data for one species of forage fish, pollock less than one year old (young pollock), from the central region only.

From some preliminary discussion of the sampling plan, the zigzag lines in Figure 1 are excluded from the analysis of this case study. Although the process of obtaining the biomass value is given below, the actual values of the study are rescaled to maintain the confidentiality of the findings of this APEX study. The data were originally standardized so that a given value of biomass is the average grams of young pollock in a one square meter (m) column between 26 m and 62 m depth in the water surveyed during a 15-second period of time. The original units were in grams per meter cubed for each one meter deep interval between 26 m and 62 m. These values have been summed over the interval 26 m to 62 m and are reported as average grams per square meter of surface area with the implication that the mean biomass is for the entire column of water whose depth is between 26 m and 62 m. Associated with each value of biomass are the longitude (X) and latitude (Y) and transect number; the longitude and latitude have been coded so that 147.7034 west and 60.3683 north gives $X = 0.0$, $Y = 0.0$; the units are in meters.

The data set is available on the floppy disk, and the file is called Case14A.txt. There are 6404 sampled sites in this region of about 40 km (25 miles) in the north–south direction and about 25.6 km (16 miles) in the east–west direction. Each site is recorded in each row of the data file. Each row consists of four entries: the X coordinate, the Y coordinate, the corresponding average biomass in one square meter of surface area (between 26 m and 62 m depth), and the transect label. Refer to Figure 2 for transect labels. A display of the sampled sites is given in Figure 2. The empty spaces in the northern and the south-western area are occupied by islands (see Figure 1).

ANALYSIS

From the extent of questions listed in a previous section and the massive amount of data collected at one point in time, it is easy to see that the scope of the analysis quickly moves beyond the goals of classroom exercises. In particular, we do not present data from two regions or times but restrict the suggested exercises to the following broad steps:
1. Plot the data and examine any systematic pattern. Where are the regions with the larger values for biomass of young pollock?
2. Check for spatial dependence and type of dependence. Are sites close together more similar than sites which are far from each other? If sites are correlated in space, how would one measure such dependence?
3. Compute the correlation over space. Fit a parametric function to the correlation (or covariance) of sites over various distances. Define a zone of influence (neighborhood) such that all sites within this zone of each other have influence on the measurement of biomass at a given site. In other words, data from correlated neighbors should be used to improve the estimate of biomass at a given site.
4. Estimation of biomass: (a) Estimate the average biomass without spatial consideration. (b) Estimate the average biomass with spatial consideration.

Exploratory Data Analysis

Plot the data and examine any systematic pattern. Where are the regions with the larger values for biomass of young pollock? Figure 3 displays the biomass in a bubble plot, where the size of the bubble represents the magnitude of biomass with the maximum scaled to 1 cm in radius. In our case study, Figure 3 displays several features.

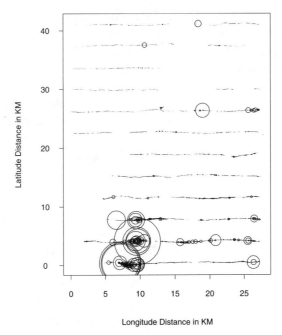

Fig. 3. *Transect biomass scales to 1 cm.*

1. Large biomass values are clustered together near the southern portion of the study area.
2. The sites within a transect are much closer than the ones across transects, and sites within a transect are likely to be correlated with each other.
3. The majority of the biomass appears to be aligned across the three southern most transects.

Readers are encouraged to plot the data in other manners to further examine the distribution of biomass in the study area.

Exploratory Look at Spatial Dependence

Check for spatial dependence and type of dependence. Are sites close together more similar than sites which are far from each other? Are sites correlated in space?

This step involves understanding the spatial dependence among these measurements by computing the sample correlogram. A sample correlogram (a spatial analogue of correlation coefficient) is defined as

$$(1) \qquad \hat{cg}(d) = \frac{1}{\#N(d)} \sum_{j \neq j^* \in N(d)} \frac{(Z(s_j) - \overline{Z})(Z(s_{j^*}) - \overline{Z})}{\sum_j^n (Z(s_j) - \overline{Z})^2 / (n-1)},$$

where $N(d)$ represents the set of sites which are d units apart in distance and $\#N(d)$ represents the number of (s_j, s_{j^*}) pairs within d distance of each other. $Z(s_j)$ represents the observed biomass at site s_j and $\overline{Z}$ is the average of all 6404 biomass values. In theory, this is easy to do, but it requires some programming. For this exercise, we selected a range of d from 0 to 8.05 km (about 5 miles) and subdivided this range into 100 intervals of size 80.5 m each. Find all pairs of sites which are within 80.5 m of each other, $\#N(80.5)$, and compute the value of the correlogram, $\hat{cg}(80.5)$, for these pairs. For each of the intervals [0 m, 80.5 m), [80.5 m, 161 m), [161 m, 241.5 m), ... , compute the above quantities and then plot $\hat{cg}(d)$ versus d. It will be a useful exercise to verify the correlogram in Figure 4 if you have programming skills. Otherwise, plot the pairs of values in the file Case14B.txt which are provided with the data diskette. Column one is distance (d) and column two is the correlogram $(\hat{cg}(d))$. The resulting plot of the correlogram for our study is in Figure 4.

The solid line in the correlogram is the zero (no correlation) line. Based on the correlogram in Figure 4, estimate the zone of influence. That is, at what distance between sites does the correlation become negligible.

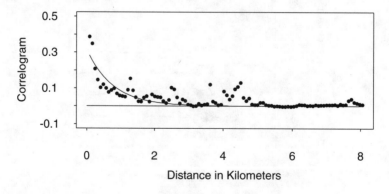

Fig. 4. *Transect biomass.*

The correlogram suggests that the degree of dependence decreases rapidly as distance increases so that sites which are approximately 1.6 km (1 mile) apart are essentially uncorrelated. According to a parametric estimation of the correlogram, the correlation at 1.6 km distance apart is nearly 0.04.

From Figures 3 and 4, one may wish to see if the spatial dependence within a transect extends beyond the distance between 2 transects. Not knowing the size and geometry of the schools, one may use the spatial dependence within a transect to extrapolate across transects. First, examine the spatial dependence for each transect in this region. Figures 5 and 6 show the correlograms for all 16 transect segments. Due to some land masses in the northern area, some transects are broken into 2 segments. Therefore, there are 16 correlograms. The last letter of the transect label indicates an intact transect with "a," an east segment transect with "e," and a west segment transect with "w." The first 3 correlograms represent the 3 southern most transects. Their spatial dependence resembles the one in Figure 4. These figures suggest that the spatial dependence pattern for the whole region is dominated by the 3 southern most transects. The correlograms from other transects either lack systematic pattern or have very small correlation values for all distances.

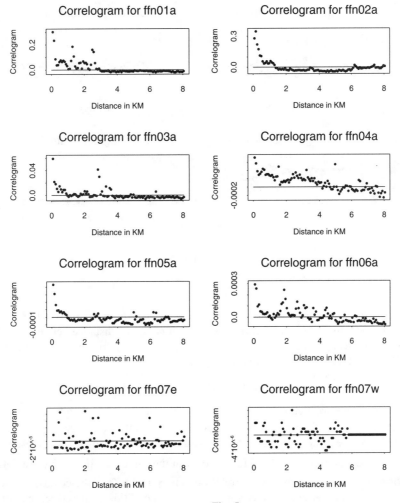

Fig. 5.

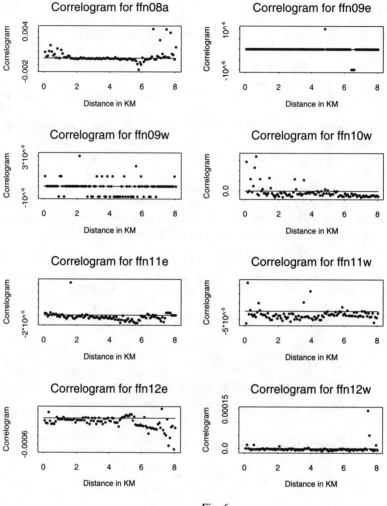

Fig. 6.

Estimation of the Spatial Correlation

Judging from the correlograms from Figures 5 and 6, it is reasonable to fit an exponential model to the decay of correlation as a function of distance between sites. This parametric relationship is needed for later analysis. The parametric function used to model the correlogram obtained from all biomass values is

$$(2) \qquad cg(d) = \beta_1 \exp(-\beta_2 d),$$

where β_1 and β_2 are positive parameters and d is the distance. Although other functional forms may be used, we found that the chosen function fits the correlogram quite well (see Figure 4). A nonlinear regression approach was used and the parameters were estimated to be $\hat{\beta}_1 = 0.314104$ and $\hat{\beta}_2 = 1.31545$ for the data in the file Case14B.txt. By the definition of autocorrelation, we mean that the correlation between two Z's is the covariance between them divided by their variance (under some stationary condition). Therefore, one may find the covariance of biomass between two sites d distance apart by multiplying the correlogram with the variance of biomass. For this data set, the sample variance estimate of biomass is 0.1684.

Estimation of Average Biomass

What are the implications of these exploratory analyses? How does spatial dependence help the biomass estimation? How would one make use of the spatial information to improve the estimation procedure? The spatial dependence can be used to estimate average and total biomass in the region and its variance in the following discussion. To compare this approach to one without spatial consideration, let us compute the total biomass and its variance assuming independence of all sites.

Estimate Average Biomass without Spatial Information

The simple "finite sampling" approach is to compute the average of all 6404 values and compute the variance and the standard deviation of the mean (standard error). The mean is 0.04867177 g/m^2 and the standard error is 0.005128495 g/m^2.

Estimation of Biomass with Spatial Information

From the discussion on exploratory analysis of spatial dependence, we have chosen a region of 1.6 km (1 mile) in radius as the zone of influence. The distance between two adjacent transects is about 3.7 km, which is larger than the observed zone of influence in the correlogram (Figure 4). It is reasonable to assume that the data between two transects are uncorrelated. A simple approach to estimate the average biomass is to compute the weighted average of the means from the transects, and then estimate the variance of the weighted average using variances from the transects. The weights for averaging are the proportion of sites for each transect in the entire region. The definition of this estimate is given as follows :

(3) $$\overline{Z}_{+} = \sum_{i}^{t} w_i \overline{Z}_i \text{ , where } w_i = n_i / n_+ \text{ , } n_+ = \sum_{i}^{t} n_i \text{ ,}$$

$\overline{Z}_i$ represents the average biomass from the ith transect, n_i represents the number of observations in the ith transect, and t represents the number of transects. The appendix gives the mean and variance of this estimate. Using this approach, the estimated mean $(\overline{Z}_+)$ biomass of young pollock per square meter is 0.04867177 g/m^2 and the standard error is 0.0050607 g/m^2. The average and standard error of biomass estimates match the ones from the previous subsection.

There are a number of models which have been proposed for analysis of spatially correlated data under the general heading of "geostatisics." Many of the early advances and development were made in the applications for geology. [Cressie, 1991] and [Ripley, 1981] are two frequently referenced books in this area. After considering several of these models for the APEX project, we chose one such model here for comparison with the estimates presented in the previous subsection and the previous paragraph.

One method to utilize the correlation of neighboring biomass values on transects is to compute a moving average at each site. The average at a site is a weighted average of its neighbors in the zone of influence where the weights are inversely proportional to the distance from the site. Following the exploratory analysis and the magnitude of the parameters from the covariance function, the zone of influence is chosen to be a neighborhood of 1.6 km (1 mile) in radius. This suggests that a correlation smaller than

0.04 is considered negligible. Using this information, let us define an estimate of biomass at each site as a weighted average of its neighbors within a 1.6 km (1 mile) radius.

$$(4) \qquad \hat{Z}(s_i) = \sum_{j \in N(i)} w_j Z(s_j),$$

where $N(i)$ is the neighborhood of sites that lie within the zone of influence from site i and w_j are weights that sum to one (they are standardized reciprocal distance weights).

$$(5) \qquad w_j = d_{i,j}^{-1} \bigg/ \sum_{k \in N(i)} d_{i,k}^{-1},$$

where $d_{i,j}$ is the distance between s_i and s_j. A farther neighbor has a smaller weight than a closer neighbor. The moment equations in the appendix indicate that $\hat{Z}(s_i)$ is unbiased for $Z(s_i)$ and that the spatial dependence enters the variance estimation through the covariance term. For interested readers, the appendix gives the equations of the mean and variance of these estimates. The contents of Case14c.txt are X Y coordinates in columns 1 and 2, the observed biomass in column 3, the estimated biomass in column 4, and the estimated variance at each site in column 5. A computer program was used to compute these estimates. Although the algorithm is relatively simple, the size of this data set makes the computation inaccessible in a classroom setting because the iterations needed to compute estimate of equation (13) are too many. In our case study,

$$\bar{\hat{Z}} = \sum_i^n \hat{Z}(s_i) / n$$

is 0.050034 g/m^2 and the standard error is 0.016919 g/m^2.

Table 1 summarizes the estimates obtained in the three methods described.

Table 1. *Estimation summary.*

	Average Biomass	Std. Error
No spatial information	0.048672 g/m^2	0.005129 g/m^2
Independent transects	0.048672 g/m^2	0.005061 g/m^2
With spatial information	0.050034 g/m^2	0.016919 g/m^2

CONCLUSIONS

When presenting this case study to the APEX team, one would use Figures 1 and 2 for background information and data description, Figure 3 for location and concentration of biomass, and Figures 4, 5, 6, and Table 1 for discussion of biomass estimation with spatial correlation.

Remarks on the estimation using spatial consideration:

1. The estimate of average biomass is slightly larger than the one given with the other two approaches. Under the constant mean assumption, all three estimates are unbiased.

2. The weighted average estimate for biomass at each site provides a "smoother" profile than the original biomass because the fluctuations across sites are averaged out by the values in the zone of influence. These values in the file Case14C.txt can be plotted for graphical presentation.

3. The variance of $\hat{Z}(s_i)$ at each site is available for graphical display or pointwise inference. The estimate ± certain multiples of standard error at each site could be plotted in three dimensions to illustrate the point estimates of biomass and the relative precision at each site.

4. The standard error of the average biomass is larger with the spatial consideration. In this case, it is more than three times as large, because data values are not independent.

5. The procedure to compute the weighted average estimates is not readily available on commercial software. The computation of the variance estimate using the equations in the appendix requires huge numbers of iterations for this data set.

REFERENCES

Cressie, N. (1991), *Statistics for Spatial Data*, John Wiley and Sons: New York.
Ripley, B.D. (1981), *Spatial Statistics*, John Wiley and Sons: New York.

BIOGRAPHIES

W. Taam received his Ph.D. in Statistics from the University of Wisconsin-Madison in 1988. He is currently an Associate Professor at Oakland University where he participates in various aspects of different outreach programs between Oakland University and neighboring industries and teaches both on- and off-campus courses. His research interests include spatial statistics and industrial statistics.

L. McDonald is the president of WEST Inc. He is a statistician and biometrician with 25 years of experience in the application of statistical methods to design, conduct, and analyze environmental and laboratory studies. Before starting WEST Inc. in 1991, Dr. McDonald held appointments in the Department of Statistics and Zoology at the University of Wyoming. His recent responsibilities on projects include design and analysis for litigation (e.g., Natural Resource Damage Assessments), wildlife surveys, habitat selection by wildlife populations, laboratory and field studies, effects of pesticides on wildlife populations, and bioassay for fresh water and marine organisms.

K. Coyle is a Research Associate at the Institute of Marine Science, University of Alaska. He has been working with quantitative acoustic systems for ten years and has participated in a number of research programs aimed at determining the horizontal and vertical distribution of prey taken by planktivorous and piscivorous seabirds. He has also been involved in research on the distribution, abundance, and species composition of zooplankton in Alaskan waters. He has participated in studies on the biology and production of benthic amphipods, preyed upon by gray whales, and has worked on taxonomy of benthic marine amphipods from around Alaska.

L. Halderson is a Professor at the School of Fisheries and Ocean Sciences at the University of Alaska. He teaches and conducts research on topics in marine fish biology. He has studied distribution and biology of fishes in major research programs in the Beaufort Sea, Bering Sea, Prince William Sound, and southeast Alaska.

CHAPTER 15

A Simplified Simulation of the Impact of Environmental Interference on Measurement Systems in an Electrical Components Testing Laboratory

David A. Fluharty, Yiqian Wang, and James D. Lynch

In the evolution of the automobile, electrical signal transmission is playing a more prominent role in automotive electronic systems. Since signal transmission involves voltages and currents in circuitry that are considerably less than for power transmission circuits, corrosion buildup in connections of such low energy circuits—referred to as dry circuits—is a considerable problem because corrosion buildup increases resistance.

To study the reliability of dry circuits in a laboratory setting, automotive engineers test connections and measure the resistance in a connection as a surrogate for connection failure. Since resistance is measured indirectly via Ohm's Law using circuit voltage and current measurements, the voltage and current measurement errors propagate through Ohm's Law[1] to affect the calculated resistance. In addition, such tests are very sensitive to external voltage sources that can be difficult, if not impossible, to control even in laboratory settings. Because these tests are performed with voltages and currents that are very small, the sensitivity of the calculated resistance to error propagation and to intermittent voltage sources are important issues. The purpose of this project is to gain insight into these issues through a simulation study.

---

[1] See [O'Malley, 1992, Chapter 2] for a basic explanation of Ohm's Law.

INTRODUCTION

The role of electrical contacts and connections in the reliability of automotive wiring—and thus the automobile as a system—is becoming of greater importance. This is because wiring is being used more and more for signal transmission rather than only for power transmission. An example of signal transmission is a small current from an exhaust sensor flowing to the engine control microprocessor. An example of power transmission is a large current flowing from the battery, through a simple switch, to the headlamp. There are numerous electrical connections along the path of current in an automobile. Increases in resistance at these connections can lead to degradation of signals. This can, in turn, cause microprocessors to issue control signals that negatively impact the performance of major vehicle systems such as the engine. Results can include hesitation, poor fuel economy, and increased emissions. Thus, the evaluation of connection systems is an important engineering task not only for automobiles, but for other products that use electronics.

Engineering these systems entails laboratory testing to simulate the "field performance" of resistance over time. In this case study, a statistical simulation is used to demonstrate the importance of understanding the laboratory environment under which such testing is conducted. Failure to do so can result in basing engineering decisions on poor data. Such decisions might result in degraded performance, increase cost, or even entail an expensive and time consuming redesign effort.

Two considerations in measuring resistance are that (i) it is calculated indirectly via Ohm's Law using measured voltages and current and (ii) the measured resistance is very sensitive to external voltage sources. Consideration (i) is related to error propagation and transmitted variation issues and considers how variation (fluctuations) and measurement errors in the voltages and the currents are transmitted to the calculated resistance through Ohm's Law. In (ii) the issue is the sensitivity of the calculated resistance to intermittent voltage that is difficult to control even in laboratory settings.

The student will use time series plots and normal probability plots for simulations of calculated resistances to study the sensitivity of the calculated resistance to (a) fluctuations in the actual currents and voltages in the circuits and (b) intermittent external voltage sources. Before the simulations can be conducted, however, a number of algebraic substitutions and manipulations—all of which relate to test procedures or realities of the electrical measurement system—are required to produce equations used in the simulation. The equations are "programmed" into a spreadsheet format to accumulate simulation results (see Table 1).

BACKGROUND INFORMATION

Several decades ago most circuits in an automobile were power circuits. These carried enough current to run various devices, e.g., headlamps, radio, turn signals, and ignition. With the advent of automotive electronics this situation has changed drastically. There are now an increasing number of microprocessors in the vehicle that require electrical signals as input. These signals are frequently from sensors that operate at very low current levels. Circuits carrying these low power signal currents are referred to as "dry circuits." For the purpose of this case study, dry circuits have voltages (V) less than 20 millivolts (20 mV) and current (I) less than 1 milliamp (1 mA). At its simplest, an automotive connection system is composed of male and female terminals that are crimped to wire leads. This sub-assembly is held in place in a plastic "hard shell" connector (see Fig. 1a). As depicted in Fig. 1b there are three contributors to the system resistance: the separable interface of the terminals (1 mΩ [one milliohm] maximum recommended for dry circuits[2]) and the crimp of the terminals to the wire (0.5 mΩ for each terminal/wire crimp). Since these resistances are in series, the overall resistance in this simplified connection system is a total of 2 mΩ (0.5 mΩ + 1.0 mΩ + 0.5 mΩ).

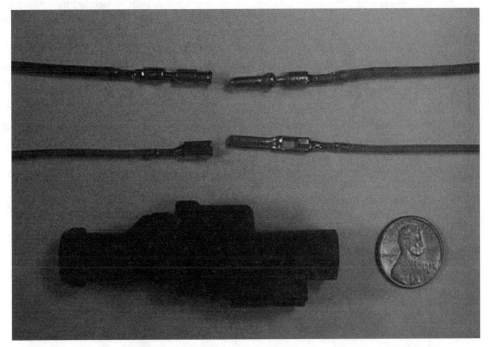

Top Row	Wire (crimped to terminal), Female Micropin Terminal, Male Micropin Terminal, Wire (crimped to terminal).
Center Row	Wire (crimped to terminal), Female Blade Terminal, Male Blade Terminal, Wire (crimped to terminal).
Bottom Row	Connected Plastic Hard Shell Connector (without wires and terminals) and US Penny for Size Comparison.

Fig. 1a. *Automotive connection systems photograph.*

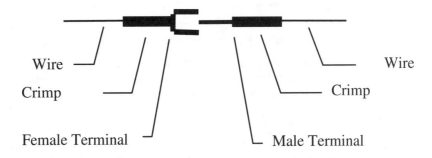

Fig. 1b. *Automotive connection system schematic.*

---

[2]Maximum resistance as required by U.S. Car, a consortium of Ford, General Motors, and Chrysler.

Chemical changes such as corrosion can occur at the molecular level in any terminal interface. These changes can result in a degradation in signal, either on an intermittent basis (possibly resulting in the problems that the technician at an auto repair facility cannot duplicate) or on an increasing basis as corrosion builds up over time. The electromotive force in a power circuit is well in excess of the breakdown voltage needed to remove such corrosion.[3] This is not the case in a dry circuit since "the voltage and/or current is too low to cause any physical change in a contact." See [J. H. Whitley, 1963].

Because automotive engineers are designing products with anticipated lives in excess of ten years and 150,000 miles, accelerated life testing is used to simulate a combination of high mileage and aging in evaluation of connection systems. One type of accelerated life testing is called combined environments testing. This may include the simultaneous application of vibration, thermal shock, and humidity in an acceleration chamber. While these stresses may be far in excess of those actually seen in the field, they are employed to achieve the activation energy necessary to excite the failure mode under investigation.[4]

Because the integrity of the signal degrades over time as resistance of the connection system increases, failure criteria frequently are specified in terms of increased resistance. According to *one* frequently used criteria, a dry circuit terminal system fails when its resistance rises 5 mΩ above its baseline resistance. For example, for the 2 mΩ system given above, the system fails if it ever reaches 7 mΩ.

To test the dry circuit connection the testing setup in Fig. 2 is used.

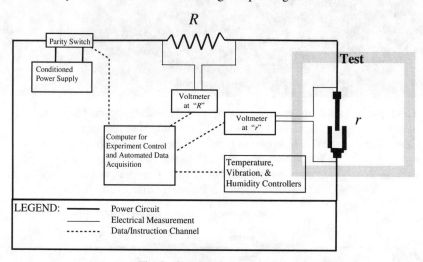

Fig. 2. *Simplified test setup (not to scale).*

In this setup circuit, two voltmeters are used to take voltage measurement from which the current and resistance are computed using Ohm's Law.[5] In Fig. 2, *r* denotes the unknown resistance of the connection system the *determination of which is the object of our test.*

---

[3]It is worth noting that there is not complete unanimity of expert opinion of the implications of contact physics for automotive dry circuits among practitioners in the field. That is one reason why testing is crucial.

[4]Establishing the correlation between combined environments testing and field performance is not a simple issue.

[5]Ohm's Law expresses the fundamental equilibrium between three elements of electricity: resistance, voltage, and current. For those not familiar with electricity, one can think of the resistance as the size of a hole through which one is trying to push a certain quantity of electricity (the current). The power with which one pushes is the voltage. Thus, for example, if resistance increases (the hole is smaller), one must either push harder (increase the voltage), or decrease the throughput (the current) to remain in equilibrium. See [O'Malley, 1992, Chapter 2].

QUESTIONS OF INTEREST

The engineering question of interest is the determination of the resistance of the connection interface, r, in the above figure. This resistance is calculated as follows:

(1) $$r = V_r / I,$$

where V_r is the voltage drop across the connection system and I is computed from the voltage reading at the other meter as follows:

(2) $$I = V_R / R.$$

Because the "resistances" r and R are in series, the current I is the same at both locations.

The actual voltages, V_r and V_R, cannot be directly observed since they are subject not only to measurement[6] error but also to other external voltage sources induced through the test and/or environmental sources. Collectively, these other voltage sources are referred to as the voltage offset,[7] which will be indicated by the symbol V_o. In addition, the current I and resistance r are never known but are also measured indirectly since they are computed using formulas (2) and (3). To avoid ambiguity, a tilde (~) over a variable will be used to distinguish the measurement or computation of a value from the actual value. Thus we have the following:

Symbol	Read as	Indicates
$\tilde{V}_r$	*V*-tilde sub small *r*	*Measured* voltage at the meter labeled little *r*
$\tilde{V}_R$	*V*-tilde sub capital *R*	*Measured* voltage at the meter labeled capital *R*
R	Capital *R*	The resistance of a known resistor, *R*. This resistor is placed in the system by the test engineer.

The voltage offset is symbolized by V_o. Thus the measured voltages at meter "r" and meter "R," V_r and V_R, respectively, are given by the following equations:

(3A) $$\tilde{V}_r = V_r + V_o,$$

(3B) $$\tilde{V}_R = V_R + V_o.$$

To get more accurate readings at r and at R, the voltages at these two locations are determined by the following procedure:
1. Send current through the circuit in one direction (called the positive).
2. Then, immediately reverse the polarity and send the current through the circuit in the opposite (negative) direction.
3. Subtract these two readings.

---

[6]An engineer never knows the "true" value of what is measured. We only have the measurement from the measurement system. The measurement system consists of the "true" value, the meter, the procedures used to calibrate the meter, the procedure used to read the meter, the person using the meter (including their training, experience, and physical skills), environmental voltage sources, and other factors. All these factors are sources of variation in the measurement system.

[7]For purposes of this exercise, we are simplifying to consider only the voltage offset. Thus factors influencing the measurement system discussed in footnote 6 are not considered. These, however, are vitally important in a laboratory to ensure that good data is obtained.

This procedure changes the parity of the circuit and cancels V_o, voltage fluctuations due to external sources such as thermal EMI (electromagnetic interference),[8] EMC/EMI (electromagnetic compatibility/electromagnetic interference) interference, temperature, and vibration if these fluctuations change slowly relative to the sampling. Note that some of these fluctuations are induced deliberately in the environmental chamber; others can come from other lab equipment; still others can come from the environment, for example, a diesel electric train traveling in the vicinity of the test laboratory.

Before we can explain how this canceling is done, we need to introduce more notation to identify the parity of the circuit. One direction of the current flow will be called positive and denoted by + and the other by a minus -. It is important to distinguish plus and minus subscripts from addition and subtraction signs.

Symbol	Symbol Explained	Indicates
$_+\tilde{V}_r$	A positive subscript before a V-tilde followed by a small r subscript.	Measured voltage at the meter labeled small r when the current has positive parity (is flowing in the "positive" direction).
$_-\tilde{V}_r$	A negative subscript before a V-tilde followed by a small r subscript.	Minus the measured voltage at the meter labeled small r when the current has negative parity (is flowing in the "negative" direction).

Thus, from Ohm's Law,[9] $V = Ir$,

(4A) $$_+\tilde{V}_R = {}_+V_r + V_o = (_+I)\ r + V_o,$$

(4B) $$_-\tilde{V}_r = {}_-V_r + V_o = (_-I)\ r + V_o,$$

(4C) $$_+\tilde{V}_R = {}_+V_R + V_o = (_+I)\ R + V_o,$$

(4D) $$_-\tilde{V}_R = {}_-V_R + V_o = (_-I)\ R + V_o.$$

You will notice that the voltage offset, V_O, does not have a parity subscript since it is external to the circuit. In (4) it has been implicitly assumed that V_o is independent of the current flow and location.[10]

The polarity in the circuit is reversed to remove the voltage offset from the measured voltage. Thus there are two readings for the voltage at r given by (4A) and (4B). Subtracting (4B) from (4A) yields the following:

(5A) $$(_+V_r + V_o) - (_-V_r + V_o) = [(_+I)\ r + V_o] - [(_-I)\ r + V_o]$$

$$= [(_+I)\ r] - [(_-I)\ r] + [V_o - V_o]$$

$$= [(_+I) - (_-I)]\ r\ .$$

---

[8]Thermal EMI is induced due to a temperature gradient along the wire.

[9]A further note on notation: $(_+I)\ r$ indicates current in the positive parity multiplied by the resistance r. The small r here is *not* a subscript as indicated by the parentheses around $_+I$.

[10]This is a major assumption. As can be seen by imposing the lines of a field on the test set up (Fig. 2), even a single electromagnetic field might have a different impact at R and r. In addition, V_o excludes accuracy of the voltmeters, which would be different with the two voltmeters used in this setup.

Direct measurements of $_+I$ and $_-I$ are not available. However, by subtracting (4D) from (4C) and rearranging the terms, an expression equivalent to $[(_+I) - (_-I)]$ is obtained:

(5B) $_+V_R - _-V_R = [(_+I) R + V_o] - [(_-I) R + V_o]$,

$= [(_+I) R] - [(_-I) R] + [V_o - V_o]$,

$= [(_+I) - (_-I)] R$,

$(_+V_R - _-V_R) / R = (_+I - _-I)$.

Substituting the expression derived in (5B) into (5A) yields the following equation (6A):

(5A) $_+V_r - _-V_r = [(_+I) - (_-I)] r$.

$_+V_r - _-V_r = [(_+V_R - _-V_R) / R] r$.

(6A) $r = R(_+V_r - _-V_r) / (_+V_R - _-V_R)$.

Equation (6A) gives the theoretical relationship between the resistances and the voltages. The resistance, r, is computed from (6A) using the measured voltages and the resistance R, which has been determined with a high degree of accuracy by the test engineer. Thus we have the following relationship which is useful in the laboratory:

(6B) $\tilde{r} = R(_+\tilde{V}_r - _-\tilde{V}_r) / (_+\tilde{V}_R - _-\tilde{V}_R)$.

Equation (6B) is the equation which relates the measurements at the voltmeters to the resistance at the separable interface of the connection system (refer to Fig. 2). To review: in equation (6B) $\tilde{r}$ is the computed resistance at r and R is the known resistance at R, which is determined by the test engineer. $_+\tilde{V}_r$ and $_-\tilde{V}_r$ are the measured voltages at r (the connection system, separable interface, placed in the test chamber) when the parity is + and -, respectively. $_+\tilde{V}_R$ and $_-\tilde{V}_R$ are the measured voltages at R when the parity is + and -, respectively.

Time Dependencies: Some Formulas

In equation (4) it was assumed that the voltage offset did not change during the polarity reversal or over time. Since the two readings are almost simultaneous during the polarity reversal this may not be totally unrealistic. It would be more realistic, though, to model fluctuations in voltage offset as a function of time. The goal of this case study is to investigate the sensitivity of the computed resistance to fluctuations of the offset. This is to determine if the voltage offset, representing environmental factors, can produce a "false positive" test result. This would be seen in a resistance rise of 5 mΩ above its baseline resistance.

For the simulations below, r and V_o will be allowed to fluctuate over time while the voltage V and the resistance R will be assumed to be constant except for changes in the polarity of V (see Fig. 2). The dependency on time and polarity in variables such as r will be denoted by $_+r(t)$ and $_-r(t)$, $t = 1,2,\ldots$. To keep time units free, the units will not be specified and sampling frequencies, rates of occurrence, etc., will be specified in terms of these units. For the purposes of tracking the polarity, it is assumed that a time unit is sufficiently long enough to obtain values at both polarities.

Key formulas relating the time dependencies and computed and measured values are given below. Note that since R and r are in series, $_+I_R(t) = {}_+I_r(t) = {}_+I(t)$ and $_-I_R(t) = {}_-I_r(t) = {}_-I(t)$. So by Ohm's Law,

(7) $_+I(t) = {}_+V/(R + {}_+r(t))$ and

$$_-I(t) = {}_-V/(R + {}_-r(t)),$$

where $R = 20$ mΩ and $V = 20$ mV. Formulas (4) and (6) with time dependencies given from (7) result in

(8A) $_+\tilde{V}_r(t) = {}_+V_r(t) + {}_+V_0(t) = {}_+I(t)\,{}_+r(t) + {}_+V_0(t)$,

(8B) $_-\tilde{V}_r(t) = {}_-V_r(t) + {}_-V_0(t) = {}_-I(t)\,{}_-r(t) + {}_-V_0(t)$,

(8C) $_+\tilde{V}_R(t) = {}_+V_R(t) + {}_+V_0(t) = {}_+I(t)\,R + {}_+V_0(t)$,

(8D) $_-\tilde{V}_R(t) = {}_-V_R(t) + {}_-V_0(t) = {}_-I(t)\,R + {}_-V_0(t)$,

(9) $\tilde{r}(t) = R\,(_+\tilde{V}_r(t) - {}_-\tilde{V}_r(t)) / (_+\tilde{V}_R(t) - {}_-\tilde{V}_R(t)).$

DATA

Instructions for the Simulation Analysis

To explore the impact of various environment factors that affect a laboratory test, the student will conduct a series of simulations. Each simulation will depend on the setting for R and V. The purpose of the simulations is to provide insight into the effect of different types of fluctuations in V_0 on false failures[11] and their relationship to the settings for R and V_0. R is a known resistance which is placed into the circuit by the test engineer. The voltage offset, V_0, is the result of thermal EMI, EMC/EMI interference, temperature, and vibration. In an effort to explore the accumulated effect of these changes, we recommend a spreadsheet approach in Table 1 (such as a MINITAB workspace).

Remember that the purpose of the laboratory test being simulated is to measure $r(t)$, the resistance of the connection system. We are interested in the behavior of this resistance over time. Specifically, we want to see if resistance ever rises 5 mΩ above its baseline resistance. All we have is a calculated value for this resistance, $\tilde{r}(t)$, in column C12 of the spreadsheet. Column C12 is calculated using equation (9). The inputs for equation (9) are the known resistance R and the measured voltages $_+\tilde{V}_r(t)$, $_-\tilde{V}_r(t)$, $_+\tilde{V}_R(t)$, and $_-\tilde{V}_R(t)$ which are in columns C8 to C11, respectively. For purposes of the simulation, the measured voltages are derived from equations (8A) through (8D). These equations require the "true" current levels $_+I(t)$ and $_-I(t)$ (which are in C6 and C7, respectively) and the "true" resistances at the interface $_+r(t)$ and $_-r(t)$ (which are in C1 and C2).

---

[11] A false failure occurs when test results indicate an interface failure (measured by one or more resistance readings of at least 5 mΩ.

Table 1a. *Simulation/calculation spreadsheet setup.*

Time Period	Simulated "True" Values		Simulated Voltage Offsets		Calculated "True" Current		Calculated Voltages Which Are Used to Calculate C12				
C1	C2	C3	C4	C5	C6	C7	C8	C9	C10	C11	C12
t	$_+r(t)$	$_-r(t)$	$_+V_o(t)$	$_-V_o(t)$	$_+I(t)$	$_-I(t)$	$_+\tilde{V}_r(t)$	$_-\tilde{V}_r(t)$	$_+\tilde{V}_R(t)$	$_-V_R(t)$	$\tilde{r}(t)$
1											
2											
⋮	⋮	⋮	⋮	⋮	⋮	⋮	⋮	⋮	⋮	⋮	⋮
Example	2.038	2.063	32.732	33.417	.908	−.907	34.581	31.548	50.882	15.287	1.704
⋮	⋮	⋮	⋮	⋮	⋮	⋮	⋮	⋮	⋮	⋮	⋮
N											

Table 1b. *Spreadsheet column descriptions.*

Column	Symbol	Variable	Use Equation	Comments
C1	t	Time subscript		
C2	$_+r(t)$	"True" resistance at r measured at time t with positive parity		Simulated per instructions in scenarios
C3	$_-r(t)$	"True" resistance at r measured at time t with negative parity		Simulated per instructions in scenarios
C4	$_+V_o(t)$	Voltage offset at time t for positive parity		Simulated per instructions in scenarios
C5	$_-V_o(t)$	Voltage offset at time t for negative parity		Simulated per instructions in scenarios
C6	$_+I(t)$	"True" current at time t for positive polarity	7	Assume $R = 20$ mΩ and $V=20$ mV
C7	$_-I(t)$	"True" current at time t for positive polarity	7	Assume $R = 20$ Ω and $V = 20$ mV
C8	$_+\tilde{V}_r(t)$		8A	Uses C6
C9	$_-\tilde{V}_r(t)$		8B	Used C7
C10	$_+\tilde{V}_R(t)$		8C	Use C6, Assume $R = 20$ mΩ
C11	$_-\tilde{V}_R(t)$		8D	Uses C7, Assume $R = 20$ mΩ
C12	$\tilde{r}(t)$		9	Assume $R = 20$ mΩ

When $_+r(t)$ and $_-r(t)$ fluctuate about a fixed level of specified resistance r, the nominal current in the circuit is defined as $V/(R+r)$. To simulate this, model $_+r(1)$, $_+r(2)$, ... and $_-r(1)$, $_-r(2)$, ... as independent normal random variables with mean $\mu = r = 2$ mΩ and a standard deviation of $\sigma = 0.05$ mΩ. Assume $R = 20$ mΩ and $V = 20$ mV. Thus the nominal current is 20 mV/(20 mΩ + 2 mΩ) = .90$\overline{909}$ mA. Theoretical considerations (Jensen's inequality) show that the mean current in the circuit (the mean of $V/(R+_+r(t))$) is strictly greater than the nominal current $.90\overline{909}$ mA. Use C6 and C7 as a test to see if you can detect this difference between the mean current and its nominal value. This difference is the so-called bias between the estimator, the average current whose mean is the mean current, and the population parameter it is estimating, the nominal current. Construct normal probability plots for columns C1 and C2 to verify that their entries are from a normal distribution. Also construct normal probability plots for C6 and C7 to see if the current can be modeled by a normal distribution when the resistances are normal.

To analyze false failures, construct time series plots of $\tilde{r}(t)$ for each of the scenarios given below. In each scenario, as above, $_+r(1), _+r(2), \ldots$ and $_-r(1), _-r(2), \ldots$ are independent normal random variables with mean $\mu = r = 2$ mΩ and a standard deviation of $\sigma = 0.05$ mΩ. To reference the false signals add a horizontal line going through 7 mΩ, the defined failure resistance. One observation through the line is a failure.

Scenario 1: V_o constant and $r(t)$ Fluctuates (see datafiles Case15A.txt and Case15B.txt)

As a baseline for comparing the impact of changes in V_o, the first scenario is just the case when $V_o(t) = V_o$ is constant over time and $r(t)$ fluctuates about a level of specified resistance r each time it is computed.

Scenario 2: Modeling Fluctuations in V_o (see datafile Case15C.txt)

2a. This is the same as Scenario 1 except that $V_o(t)$ fluctuates about a level V_o. To model this fluctuation, let $_+V(1), _+V(2), \ldots$ and $_-V(1), _-V(2), \ldots$ be independent normal random variables with mean $\mu = V_o$ and $\sigma = V_o/9$, where $V_o = 1, 10, 20,$ and 30 mV. (Note that the coefficient of variation $\sigma/\mu = 1/9$, or in terms of signal to noise, $\mu/\sigma = 9$. Since almost all of a normal distribution's values (99.73%) are within 3 standard deviations of its mean, one interpretation of $\sigma/\mu = 1/9$ is that the error in measurement relative to V_o is at most about $3\sigma/\mu = 1/3 = 33\%$.) At what values of V_o do sporadic false failures occur? When do the false failures become chronic?

Scenario 3: Modeling Occurrences of Disruptions in V_o

3a. Use a Poisson process to model environmental factors such as changes in temperature due a door opening or flipping a light switch that may effect a significant change in V_o, say an increase of 20 mV. To do this, assume that there are on the average 2 per 100 time units. Generate a Poisson random variable P with mean equal to $2*(N/100)$, where N is the number of rows in the spreadsheet. Randomly select P numbers without replacement from 1, 2, ..., N. These will correspond to the P rows in column C13 where one hundreds will be placed. The remaining $N-P$ rows of C13 will be filled with zeroes. Add this column to the C4 and store in C14 to simulate this type of disruption.

3b. To simulate a single major disruption of V_o, such as a train passing (which actually was a problem), randomly select a single number between 1 and N. Add 50 mV to this row and the next 9 entries in C4 and store in C15.

ANALYSIS

The graph in Fig. 3 is typical of a time series graph of $\tilde{r}(t)$. Note that there are at least four instances where resistance is over 7 mΩ, that is, more than 5 mΩ above the allowable 2 mΩ (see Background Information). These are "false failures."

(Note: There are resistance values below 0. This is not possible in practice and could have been prevented with more sophisticated programming in the simulation, equating all negative values to zero.)

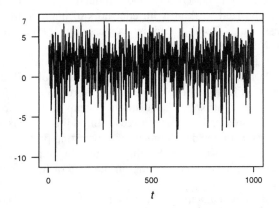

Fig. 3. *Time series graph of calculated resistance (includes "false failures").*

INSTRUCTIONS FOR PRESENTATION OF RESULTS

The student should perform the indicated simulations. Normal probability plots will give an indication of the distributional behavior of the input resistances to the voltages, currents, and calculated resistances. Time series plots can be used to determine the prevalence of false signals of connection failure when the calculated resistance exceeds 7 mΩ.

REFERENCES

M. Antler (1982), *Field Studies of Contact Materials: Contact Resistance Behavior of Some Base and Nobel Metals*, IEEE Transactions on Components, Hybrids, and Manufacturing Technology, Vol. CHMT-5, No. 3, September, pp. 301–307.

R.P. Heinrichs (1983), *Considerations for the Contact Interface In Tomorrow's Automotive Electronics*, paper presented at the Society of Automotive Engineers (SAE) International Congress and Exposition (SAE Publication 930422), Society of Automotive Engineers, Warrendale, PA.

J. O'Malley (1992), *Basic Circuit Analysis* (Chapter 2 for a basic explanation of Ohm's Law), Schaum's Outline Series, McGraw–Hill, Inc., NY.

J. H. Whitley (1963), *Contacts and Dry Circuits,* paper presented at an Invitational Symposium, Montreal, Canada, AMP Incorporated, Harrisburg, PA.

J. H. Whitley (1980), *Connector Surface Plating: A Discussion of Gold and the Alternatives*, AMP Incorporated, Harrisburg, PA.

K-I. Yasuda, S. Umemura, and T. Aoki (1987), *Degredation Mechanisms in Tin- and Gold-Plated Connector Contacts*, IEEE Transactions on Components, Hybrids, and Manufacturing Technology, Vol. CHMT-10, No. 3, September, pp. 456–462.

BIOGRAPHIES

David Fluharty (MA and MBA, University of Chicago) is the Manager of Reliability Engineering at the Dearborn Michigan office of AFL Automotive. He has been with AFL, a $1.4 billion automotive parts and telecommunications company, for eight years. In addition to his primary responsibilities in developing and managing systems to improve product reliability, he consults in areas of statistics, quality, and strategic planning. Prior to AFL he spent eleven years at Ford Motor Company in a financial analysis and statistical consulting position. He is active in the American Statistical Association, having served as the first Chair of the ASA Section on Quality and Productivity. More recently his professional efforts have emphasized working with educators to improve the teaching of statistics in the K–12 classroom. To this end he is pursuing a Ph.D. in Education Evaluation and Research at Wayne State University.

Yiqian (E-Chan) Wang (Ph.D., Wayne State University) is the Senior Test & Reliability Engineer at the Dearborn Michigan Reliability Laboratory of AFL Automotive. His primary responsibilities at AFL are to develop and improve reliability testing methods and techniques. He designs test control/data acquisition systems that perform tests that meet automotive industry standards and customer needs. In addition to his work at the Reliability Laboratory, he consults on the development of new testing protocols. He served as the Associate Director of the 21st Annual Reliability Testing Institute at the University of Arizona. Prior to joining AFL he worked in R&D in the area of his primary research interest, Thermal Wave Imaging.

J. D. Lynch (Ph.D., Florida State University), Professor of Statistics and Director of the Center for Reliability and Quality Sciences at the University of South Carolina, has taught and consulted for a number of companies on various quality and reliability projects. These industrial consulting and training experiences have been influential in the revamping of the department's engineering statistics course and its quality control course. He has published numerous papers in the area of reliability, and his research has been funded in the past by the National Science Foundation and the Army Research Office. His current research, on the reliability of complex systems with dependent components and complex materials, is funded by the National Science Foundation.

CHAPTER 16

CEREBRAL BLOOD FLOW CYCLING: ANESTHESIA AND ARTERIAL BLOOD PRESSURE

Michael H. Kutner, Kirk A. Easley, Stephen C. Jones, and G. Rex Bryce

Cerebral blood flow (CBF) sustains mental activity and thought. Oscillations of CBF at a frequency of 6 per minute, termed CBF cycling, have been suspected of being dependent on the type of anesthesia [Clark Jr., Misrahy, and Fox, 1958; Hundley et al., 1988]. Thus, we investigated the effects on CBF cycling using different anesthetics [Jones et al., 1995].

INTRODUCTION

CBF is important because it sustains mental activity and thought. The research question asked here was whether CBF cycling is influenced by the type of anesthesia. Because cycling is enhanced at lower arterial pressures, blood was withdrawn (exsanguinated) in all experimental animals to reduce their arterial pressure. Analysis of covariance is used to explore whether cycling characteristics, amplitude, and frequency differ by type of anesthesia while controlling for the amount of blood pressure change induced by exsanguination.

BACKGROUND INFORMATION

CBF is important because it sustains neuronal activity, the supposed basis of mental activity and thought [Chien, 1985]. As various regions of the brain are activated by sensory or motor demands, the level of blood flow adjusts regionally in a dynamic fashion to support the local changes in neuronal activity [Lindauer, Dirnagl, and Villringer, 1993].

Fluctuations or oscillations of CBF and their relation to the rate of oxygen use by the brain have been noted by many workers [Clark Jr., Misrahy, and Fox, 1958; Vern et al., 1988], but generally they have only been reported as secondary results and usually only in a small fraction of the subjects studied. Because these oscillations occur with a dominant frequency near 0.1 Hz, or 6/min, and often have a high amplitude that can approach 15% of the mean value of CBF, they have intrigued many who have sought to understand their physiological significance and possible relation to pathology [Jöbsis et al., 1977; Mayevsky and Ziv, 1991].

Laser Doppler flowmetry [Stern, 1975; Frerichs and Feuerstein, 1990] provides a method of monitoring rapid changes of CBF and is based on the Doppler shift that moving

Laser Doppler flowmetry [Stern, 1975; Frerichs and Feuerstein, 1990] provides a method of monitoring rapid changes of CBF and is based on the Doppler shift that moving red blood cells impart to a beam of laser light. Both this shift, which is proportional to red blood cell velocity, and the amount of the reflected laser light, which is proportional to the red cell mass, are used to provide an index of CBF that has been shown to be proportional to relative changes in CBF [Haberl et al., 1989]. However, laser Doppler flowmetry cannot provide an absolute measure of CBF.

The advent of laser Doppler flowmetry has resulted in several investigations of CBF cycling [Hudetz, Roman, and Harder, 1992; Morita-Tsuzuki, Bouskela, and Hardebo, 1992]. Since laser Doppler flowmetry is sensitive to red blood cell flow, these observations are, in a strict sense, observations of red blood cell cycling. Based on our preliminary experiments [Williams et al., 1992] and other recent data [Hudetz, Roman, and Harder, 1992; Morita-Tsuzuki, Bouskela, and Hardebo, 1992], cycling is increased at lower arterial pressures, so we chose to lower arterial pressures using exsanguination as a mechanism to increase the occurrence of cycling.

The fast Fourier transform was used to characterize the CBF oscillations by providing an estimate of the dominant frequency and relative amplitude. An example of this analysis is shown in Fig. 1. The plot of the original data in Fig. 1 (Panel A) shows the arterial blood pressure tracing and the simultaneously recorded CBF, showing a transient drop just after the steep decrease in arterial blood pressure, followed by CBF oscillations. In Fig. 1 (Panel B), the fast Fourier transform is shown of the 64 second epoch marked by arrows in Fig.1 (Panel A), with a high amplitude peak at a dominant frequency of 0.1 Hz.

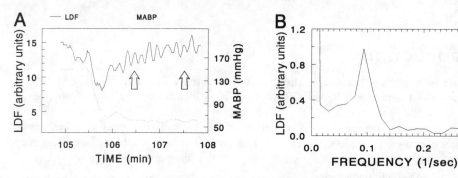

Fig. 1 *(Panel A). Dramatic initiation of cerebral blood flow (CBF) cycling (solid line) when the MABP (dashed line) is dropped from 155 to 60 mmHg (Table 1, epoch 29). Seventy-one seconds after the blood withdrawal, the 64 second epoch from which the fast Fourier transform spectrum is derived is marked with arrows.*

(Panel B). The fast Fourier transform derived spectrum showing the dominant frequency of 0.094 Hz and amplitude of 7.1%. The amplitude is calculated from the amplitude at zero frequency or the mean amplitude, 13.8 arbitrary units (not shown), and the amplitude at the dominant frequency, 0.98 arbitrary units (100*0.98/13.8 = 7.1%). Reprinted with permission from Jones, Williams, Shea, Easley, and Wei, Cortical cerebral blood flow cycling: Anesthesia and arterial blood pressure, *Amer. J. Physiol.,* 268: H569–H575.

QUESTIONS OF INTEREST

Does cerebral blood flow cycling as characterized by amplitude and frequency vary depending upon the type of anesthesia adjusted for blood pressure change?

DATA

Fifteen Sprague–Dawley rats were anesthetized with either Pentobarbital sodium, ($n =$ 5, 40–50 mg/kg IV), α-Chloralose ($n = 5$, 60 mg/kg IV), or Halothane ($n = 5$, 0.5–1.0% by inhalation). Mean arterial blood pressure (MABP) and laser Doppler flow (LDF) were recorded continuously [Jones et al., 1995]. MABP was transiently and repeatedly lowered by rapid withdrawal of blood from the femoral arterial catheter until cycling was noted. The blood was then reinfused and each animal had from 1 to 4 individual pressure drops. Thirty-six instances of cycling were observed in all 15 animals after these rapid arterial pressure drops. Epochs were chosen from the recordings to purposely represent cycling with the highest amplitude. For the purposes of this case study, these epochs were considered to be independent experimental units. Table 1 provides the raw data from each epoch including a description of each variable.

Table 1. *Raw data from fast Fourier transform (FFT) analysis of CBF.*

Variable Name	Description
Cycling Parameters:	
Amplitude	FFT-determined amplitude (%)
Frequency	FFT-determined frequency (Hz)
Physiological Variables:	
Blood Pressure	Mean arterial pressure (mmHg) just after blood withdrawal during each epoch of cycling
Blood Pressure Change	The change in blood pressure [$\%\Delta MABP = 100(MABP_{cyc} - MABP_{pre})/MABP_{pre}$], where $MABP_{pre}$ and $MABP_{cyc}$ are measured before and after the blood withdrawal.
Treatment:	
Anesthetic	P = Pentobarbital H = Halothane C = α-Chloralose

Epoch	Anesthetic	Blood Pressure	Blood Pressure Change	Amplitude	Frequency
1	C	65	−45.8	12.9	.089
2	C	55	−64.5	10.3	.081
3	H	68	−51.4	5.1	.102
4	H	76	−30.9	3.9	.078
5	H	65	−53.6	9.8	.063
6	H	65	−35.0	5.8	.078
7	P	55	−57.7	11.8	.070
8	P	60	−55.6	7.3	.094
9	P	60	−47.8	3.4	.094
10	P	70	−46.2	8.2	.094
11	P	72	−24.2	3.7	.125
12	C	50	−64.3	7.5	.086

Epoch	Anesthetic	Blood Pressure	Blood Pressure Change	Amplitude	Frequency
13	C	52	−56.7	7.0	.125
14	H	75	−28.6	4.5	.094
15	C	60	−60.0	4.5	.125
16	C	60	−50.8	14.6	.093
17	C	75	−33.0	6.8	.110
18	H	58	−49.6	6.0	.096
19	H	67	−44.2	9.3	.078
20	P	57	−61.2	5.9	.078
21	P	52	−44.7	4.7	.102
22	P	75	−45.7	5.4	.109
23	P	75	−16.7	2.1	.125
24	C	70	−50.0	6.4	.133
25	C	70	−39.1	4.1	.109
26	H	67	−44.2	3.7	.063
27	C	72	−44.6	6.8	.070
28	H	52	−55.6	6.1	.094
29	P	57	−63.2	7.1	.094
30	P	70	−41.7	2.0	.094
31	C	65	−43.5	10.7	.102
32	C	70	−30.0	3.7	.109
33	C	82	−29.3	6.1	.078
34	C	58	−53.6	11.4	.086
35	H	60	−36.8	4.7	.078
36	C	85	−34.6	5.3	.078

ANALYSIS

Analysis of covariance can be used to address the question of whether cerebral blood flow cycling differs depending upon the type of anesthesia. Two examples will be given to demonstrate the use of analysis of covariance.

Example One

The data consist of 36 experimental units ($N = 36$ epochs) with cycling amplitude (%) as the dependent variable and blood pressure change (%) as the covariate for each of the 3 anesthetics with n_i units per anesthetic ($n_1 = 11$ for Pentobarbital, $n_2 = 15$ for α-Chloralose, and $n_3 = 10$ for the Halothane (see Table 1 and Fig. 2a)). Let Y_{ij} (amplitude) denote the jth observation for the ith treatment (anesthetic) and X_{ij} denote the covariate (blood pressure change) corresponding to the (i,j)th experimental unit (epoch). We assume that the mean of Y_{ij} can be expressed as a linear function of the covariate with possibly different slopes and intercepts required for each anesthetic. An important point to note is that the mean of Y depends on the value of the covariate as well as on the particular anesthetic from which the observation was obtained.

The single-factor analysis of covariance model with one linear covariate is

(1) $Y_{ij} = \mu_i + \beta_j (X_{ij} - \overline{X}_{..}) + \varepsilon_{ij}$ $i = 1, K, t$ $j = 1, K\, n_i$

where $\overline{X}_{..}$ is the overall mean of the X_{ij}'s. Here we will assume that error terms are independently, identically, normally distributed with mean zero and variance σ^2, i.e., $\varepsilon_{ij} \sim iid \ N(0, \sigma^2)$. The required SAS statements to fit this model are provided on page 137.

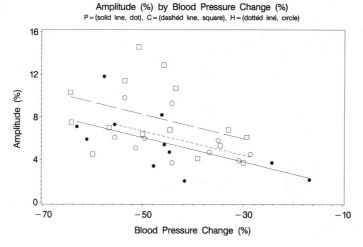

Fig. 2a.

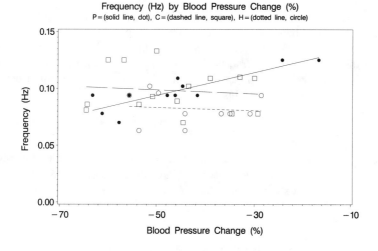

Fig. 2b.

This model has seven parameters $(2t+1)$, three intercepts μ_i (when $X_{ij} = \overline{X}_{..}$), and three slopes β_i and σ^2, the variance of each experimental unit. The least squares estimates of the parameters μ_i and β_i can be obtained by minimizing the residual sum of squares in (2):

$$(2) \qquad SSRes = \sum\sum [Y_{ij} - \mu_i - \beta_i(X_{ij} - \overline{X}_{..})]^2 .$$

The least squares estimator of each β_i is the same as the estimator obtained for a simple linear regression model. The residual sum of squares for the ith treatment is based on $n_i - 2$ degrees of freedom and is

$$SSRes_i = \sum_{j=1}^{n_i} [Y_{ij} - \mu_i - \beta_i(X_{ij} - \overline{X}_{..})]^2 \; .$$

The residual sum of squares for model (1) can be obtained by pooling residual sums of the squares for each of the three models to obtain

$$SSRes = \sum_{i=1}^{3} SSRes_i \; .$$

In general, the residual sum of squares is based on $(N - t(k+1))$ degrees of freedom, where

$$N = \sum_{i=1}^{t} n_i$$

and k = the number of covariates. The best estimate of the variance is $\hat{\sigma}^2 = SSRes/(N - t(k+1))$. The basic hypotheses most commonly tested about the parameters of the analysis of covariance model are discussed in the next section.

Testing Hypotheses about the Slopes: Example One

1. Does the mean of amplitude given the blood pressure change depend linearly on the covariate? This question can be answered statistically by testing the hypothesis

$$H_{01} : \beta_1 = \beta_2 = \beta_3 = 0 \qquad \text{vs.} \qquad H_{a1} : (\text{not } H_{01}) \; .$$

The null hypothesis states that none of the anesthetics have means which depend on the value of blood pressure change (SAS code provided on page 137). A model comparison method provides a way of obtaining the desired test statistic. The model restricted by the conditions of the null hypothesis, H_{01}, is $Y_{ij} = \mu_i + \varepsilon_{ij}$, i = 1, 2, 3 and j = 1, ...n_j. This model is the usual analysis of variance model for the one-way treatment structure. The residual sum of squares for this model is

$$SSRes(H_{01}) = \sum_{i=1}^{3} \sum_{j=1}^{n_i} (Y_{ij} - \overline{Y}_{i.})^2 \; ,$$

which is based on $N - t$ degrees of freedom. The sums of squares due to deviations from H_{01}, denoted by SSH_{01}, is $SSH_{01} = SSRes\,(H_{01}) - SSRes$, which is based on $(N - t) - (N - 2t) = 3$ degrees of freedom. A test statistic for testing H_{01} versus H_{a1} is

$$F_{H_{01}} = \frac{SSH_{01}/t}{\hat{\sigma}^2} \; .$$

When H_{01} is true the sampling distribution of $F_{H_{01}}$ is an F distribution with t and $N - 2t$ degrees of freedom. By processing the SAS code provided on page 137, we obtain a statistically significant F statistic (F = 3.04, p = 0.044). We therefore reject H01 and conclude that the mean of amplitude does depend on the value of blood pressure change for at least one of the anesthetics. Please verify this result by calculating $F_{H_{01}}$ for the data set provided. What is the best estimate of the variance?

2. Do the cycling amplitude means of the anesthetics depend on blood pressure change differently (nonparallel lines)?

To answer this question, we test for homogeneity of the slopes. (SAS code provided at the bottom of page 137.) The appropriate null hypothesis can be stated as H_{02}: $\beta_1 = \beta_2 = \beta_3$

$= \beta$ versus H_{a2} (not H_{02}), where β is unspecified and represents the common slope of the three regression lines. The model which satisfies the conditions of H_{02} is

(4) $Y_{ij} = \mu_i + \beta(X_{ij} - \overline{X}_{..}) + \varepsilon_{ij}, \quad i = 1,\ldots,t \quad j = 1,\ldots,n_i,$

which represents three parallel lines each with slope β and intercepts μ_1, μ_2, μ_3 when $X_{ij} = \overline{X}_{..}$. The residual sum of squares for the model is

$$SSRes(H_{02}) = \sum_{i=1}^{3}\sum_{j=1}^{n_i}[Y_{ij} - \hat{\mu}_i - \hat{\beta}(X_{ij} - \overline{X}_{..})]^2 ,$$

where $\hat{\mu}_i$ and β denote the least squares estimators of the model parameters. The residual sum of squares for this model is based on $N - t - 1$ degrees of freedom. The sum of squares due to deviations from H_{02} is $SSH_{02} = SSRes\ (H_{02}) - SSRes$. The statistic used to test H_{02} is

$$F_{H_{02}} = \frac{SSH_{02}/(t-1)}{\hat{\sigma}^2} ,$$

which has sampling distribution F with $t - 1$ and $N - 2t$ degrees of freedom. The F statistic is not significant ($p = 0.93$) for this example, allowing us to conclude that the lines are parallel (equal slopes). The estimated common slope is -0.114 (standard error $= 0.0370$), which is significantly different from zero ($p = 0.004$; see Table 2). Obtain a 95% confidence interval for the common slope estimate.

3. Using the parallel lines or equal slopes model, are the distances between the regression lines different from zero? (SAS code provided on page 138.)

Since the equal slopes model is appropriate for this example, a property of parallel lines is that they are the same distance apart for every value of X. Thus, the distances between these lines can be measured by comparing the intercepts of the lines, i.e., when $X_{ij} = \overline{X}_{..}$. The hypothesis to be tested is that the adjusted means or distances between the lines are equal, which is equivalent to testing the hypothesis that the intercepts are equal:

$$H_{03} : \mu_1 = \mu_2 = \mu_3 = \mu \qquad \text{vs.} \qquad H_{a3} : (\text{not } H_{03}),$$

where μ is unspecified. The model to describe the mean amplitude as a function of the covariate and the treatments is assumed to be of the form $Y_{ij} = \mu_i + \beta(X_{ij} - \overline{X}_{..}) + \varepsilon_{ij}$. The residual sum of squares for the model is $SSRes\ (H_{02})$ given previously. This model, restricted by the conditions of H_{03}, is $Y_{ij} = \mu + \beta(X_{ij} - \overline{X}_{..}) + \varepsilon_{ij}$, which is a simple linear regression model fit to all of the data. The corresponding residual sum of squares is

$$SSRes(H_{03}) = \sum_{i=1}^{3}\sum_{j=1}^{n_i}[Y_{ij} - \hat{\mu} - \hat{\beta}(X_{ij} - \overline{X}_{..})]^2 ,$$

where $\hat{\mu}$ and β are least squares estimators from the model. The residual sum of squares for H_{03} is based on $N - 2$ degrees of freedom (DF). The sum of squares due to deviations from H_{03}, given that H_{02} is true, is $SSH_{03} = SSRes(H_{03}) - SSRes(H_{02})$, which is based on $DF(SSRes(H_{03})) - DF(SSRes(H_{02})) = 2$. The appropriate test statistic is

$$F_{H_{03}} = \frac{SS_{H_{03}}/2}{SSRes(H_{02})/32} .$$

The calculated F statistic is 2.5 ($p = 0.10$), suggesting that distances between pairs of lines may not be different from zero or that the intercepts are equal (Table 2). In other words, the cycling amplitude is related to blood pressure change but the intercepts are similar and thus we conclude there are no differences between anesthetics.

4.   Use fitted regression model (4) equations to calculate the adjusted amplitude means at $X = -45.4\%$ for each anesthetic. (LSMEANS SAS statement on page 138 provides these estimates.) Compare these values to the unadjusted mean values.

$$\hat{Y}_C = \hat{\mu}_1 + \hat{\beta}(X - \overline{X}_{..}) = 7.73 - 0.114(X_{ij} - \overline{X}_{..}),$$

$$\hat{Y}_P = \hat{\mu}_2 + \hat{\beta}(X - \overline{X}_{..}) = 5.55 - 0.114(X_{ij} - \overline{X}_{..}),$$

$$\hat{Y}_H = \hat{\mu}_3 + \hat{\beta}(X - \overline{X}_{..}) = 6.17 - 0.114(X_{ij} - \overline{X}_{..}),$$

where $\overline{X}_{..} = -45.4$ = the overall mean blood pressure change (%) for all three anesthetics.

5.   Is the adjusted mean for cycling amplitude under α-Chloralose anesthesia significantly different from the adjusted mean for cycling amplitude under Pentobarbital? (Provided by the first ESTIMATE statement on page 138.)

Testing Hypotheses About Slopes: Example Two

1.   Use analysis of covariance to examine the relationship between cycling frequency (Hz) and blood pressure change (%) for the three anesthetics.

As conveyed by Fig. 2b, the equal slopes model ($\beta_1 = \beta_2 = \beta_3 = \beta$) does not appear appropriate for the relationship between cycling frequency and blood pressure change. In this case, model (1) is necessary to adequately describe the data (see SAS code on page 139). For the following fitted regression equations, calculate the adjusted mean cycling frequency for each anesthetic at $\overline{X}_{..} = -45.4\%$:

$$\hat{Y}_C = \hat{\mu}_1 + \hat{\beta}_1(X - \overline{X}_{..}) = 0.0981 - 0.000173\,(X_{ij} - \overline{X}_{..}),$$

$$\hat{Y}_P = \hat{\mu}_2 + \hat{\beta}_2(X - \overline{X}_{..}) = 0.0986 + 0.001002\,(X_{ij} - \overline{X}_{..}),$$

$$\hat{Y}_H = \hat{\mu}_3 + \hat{\beta}_3(X - \overline{X}_{..}) = 0.0827 - 0.000136\,(X_{ij} - \overline{X}_{..}),$$

where $\overline{X}_{..} = -45.4\%$ = the overall mean blood pressure change (%) for all three anesthetics. The adjusted means are provided using the SAS ESTIMATE statement (see page 139, ESTIMATE statements 10, 11, and 12).

2.   Test the equality of the distances between the lines at a blood pressure change of $X_1 = -30\%$ (see page 139, ESTIMATE statements 18, 19, and 20).

The hypothesis can be answered by testing $H_{04|x_1=-30} : \mu_{1|x=x_1} = \mu_{2|x=x_1} = \mu_{3|x=x_1}$ or equivalently as $Y_{ij} = \mu_{i|x=x_i} + \beta_i(X_{ij} - \overline{X}_{..}) + \varepsilon_{ij}$. The model restricted by $H_{04|x=x_1}$ is

$Y_{ij} = \mu_{i|x=x_i} + \beta_i(X_{ij} - \overline{X}_{..}) + \varepsilon_{ij}$. This model is identical to model (1) with X_{ij} set equal to X_1. The corresponding residual sum of squares is

$$SSRes(H_{04|x=x_1}) = \sum_{i=1}^{t} \sum_{j=1}^{n_i} (Y_{ij} - \hat{\mu}_{i|x=x_1} - \hat{\beta}_i(X - \overline{X}_{..}))^2,$$

which is based on $N - t - 1$ degrees of freedom. The sum of squares due to deviations from $H_{04|x=x_1}$ is $SS(H_{04|x=x_1}) = SSRes(H_{04|x=x_1}) - SSRes$ with $t - 1$ degrees of freedom. The resulting test statistic is

$$F_{H_{04|x=x_i}} = \frac{(SS_{H_{04|x=x_1}})/(t-1)}{SSRes/(N-2t)}.$$

The results (Table 4) suggest that the frequency response differs significantly for one of the anesthetics compared to pentobarbital at a mean blood pressure change of -30%. (Which one? See ESTIMATE statement 20 on page 139.) Are the results similar at the overall mean for blood pressure change (mean = -45.4%)?

ANALYSIS OF COVARIANCE COMPUTATIONS USING THE SAS SYSTEM

SAS can be used to compute the various estimators and tests of hypotheses discussed in the previous sections. The SAS statements required for each part of the analysis discussed above are presented in this section.

All the following models will be fit assuming that the data from Table 1 were read in by the following SAS statements:

```
DATA ANCOVA;
INPUT EPOCH ANES$ BP BPCHANGE AMP FREQ;
DATALINES;
```

The required SAS statements needed to fit model (1) for example one are

```
PROC GLM;
CLASSES ANES;
MODEL AMP = ANES BPCHANGE*ANES/NOINT SOLUTION;
```

The term ANES with the no-intercept option (NOINT) enables one to obtain the intercept estimates for each anesthetic. The term BPCHANGE*ANES generates the part of the design matrix corresponding to the slopes. The option SOLUTION is used so that the estimates and their standard errors are printed. The sum of squares corresponding to ERROR is $SSRes$ and the MEAN SQUARE ERROR is $\hat{\sigma}^2$. The type III sum of squares corresponding to BPCHANGE*ANES tests H_{01}.

To test the equality of slopes hypothesis (H_{02}), the required SAS statements are

```
PROC GLM;
CLASSES ANES;
MODEL AMP = ANES BPCHANGE BPCHANGE*ANES/E3 SOLUTION;
```

The type III sums of squares corresponding to BPCHANGE*ANES tests H_{02}. The type III sum of squares corresponding to BPCHANGE tests that the average value of all slopes is zero, and the type III sum of squares corresponding to ANES tests H_{03} given that the slopes are unequal. By including BPCHANGE and/or removing the NOINT option, the model becomes singular and thus the set of estimates obtained is just one of the many least squares solutions. Option E3 provides a list of estimable functions that will allow us to verify that the above type III sum of squares do in fact test the indicated hypotheses. Since we fail to reject H_{02}, then we should fit the parallel lines or equal slopes model. The appropriate SAS statements are

```
PROC GLM;
CLASSES ANES;
MODEL AMP = ANES BPCHANGE/SOLUTION;
LSMEANS ANES/STDERR PDIFF;
Estimate 'C vs. P'  Anes 1 0 -1;
Estimate 'C Intercept' intercept 1  anes 1 0 0;
Estimate 'C Adj. Mean'  intercept 1   anes 1 0 0 bpchange
-45.4;
```

The results from fitting this model can be found in Table 2. The type III sum of squares and resulting F test corresponding to ANES tests that the distance between the lines is zero given that the parallel lines model is adequate to describe the data. The LSMEANS statement provides the adjusted amplitude mean given BPCHANGE = −45.4% for each anesthetic. The option STDERR provides the corresponding standard error of the adjusted means and PDIFF provides significance levels for t-tests of $\mu_{Y_{i|x=\bar{x}_\cdot}} = \mu_{Y_{i'|x=\bar{x}_\cdot}}$ for each pair of adjusted means. The first ESTIMATE statement also produces a computed t-value and its significance level comparing the intercept estimates for Pentobarbital and α-Chloralose.

Table 2. *Equal slopes model for example one. General linear models procedure.*

Dependent Variable: AMP

Source	DF	Sum of Squares	Mean Square	F Value	Pr > F
Model	3	104.49280047	34.83093349	5.17	0.0050
Error	32	215.49942175	6.73435693		
Corrected Total	35	319.99222222			

R-Square	C.V.	Root MSE	AMP Mean
0.326548	39.15436	2.5950640	6.6277778

Source	DF	Type III SS	Mean Square	F Value	Pr > F
ANES	2	33.13371493	16.56685746	2.46	0.1014
BPCHANGE	2	64.15891158	64.15891158	9.53	0.0042

Parameter		Estimate	T for H0: Parameter = 0	Pr > |T|	Std Error of Estimate
INTERCEPT		0.363265134 B	0.19	0.8471	1.86833483
ANES	C	2.185276224 B	2.12	0.0418	1.03052670
	H	0.620058522 B	0.54	0.5899	1.13889620
	P	0.000000000 B	.	.	.
BPCHANGE		−0.114135295	−3.09	0.0042	0.03697767

Least Squares Means				
ANES	AMP LSMEAN	Std Err LSMEAN	Pr > |T| H0:LSMEAN = 0	LSMEAN Number
C	7.73028376	0.67164352	0.0001	1
H	6.16506606	0.82545587	0.0001	2
P	5.54500754	0.78264404	0.0001	3

Pr > |T| H0: LSMEAN(i) = LSMEAN(j)			
i/j	1	2	3
1	.	0.1525	0.0418
2	0.1525	.	0.5899
3	0.0418	0.5899	.

NOTE: To ensure overall protection level, only probabilities associated with preplanned comparisons should be used.

Parameter	Estimate	T for HO: Parameter = 0	Pr > |T|	Std Error of Estimate
C vs. P	2.18527622	2.12	0.0418	1.03052670
C Intercept	2.54854136	1.38	0.1780	1.85068527
C Adj Mean	7.73028376	11.51	0.0001	0.67164352

The SAS statements necessary to test H_{01} (i.e., does the mean frequency given the blood pressure change depend linearly on the covariate?) for example two and estimate several linear combinations of the parameters are

```
PROC GLM; CLASS ANES;
MODEL FREQ = ANES BPCHANGE*ANES/NOINT;
Estimate 'C Intercept' Anes 1 0 0 ;
Estimate 'H Intercept' Anes 0 1 0 ;
Estimate 'P Intercept' Anes 0 0 1 ;
Estimate 'C Slope'     Bpchange*Anes 1  0  0;
Estimate 'H Slope'     Bpchange*Anes 0  1  0;
Estimate 'P Slope'     Bpchange*Anes 0  0  1;
Estimate 'C-H Slope'   Bpchange*Anes 1 -1  0;
Estimate 'C-P Slope'   Bpchange*Anes 1  0 -1;
Estimate 'H-P Slope'   Bpchange*Anes 0  1 -1;
Estimate 'C   AT -45%' Anes 1  0  0 Bpchange*Anes -45.4     0      0;
Estimate 'H   AT -45%' Anes 0  1  0 Bpchange*Anes     0 -45.4      0;
Estimate 'P   AT -45%' Anes 0  0  1 Bpchange*Anes     0     0  -45.4;
Estimate 'C-H AT -45%' Anes 1 -1  0 Bpchange*Anes -45.4  45.4      0;
Estimate 'H-P AT -45%' Anes 0  1 -1 Bpchange*Anes     0 -45.4   45.4;
Estimate 'C   AT -30%' Anes 1  0  0 Bpchange*Anes   -30     0      0;
Estimate 'H   AT -30%' Anes 0  1  0 Bpchange*Anes     0   -30      0;
Estimate 'P   AT -30%' Anes 0  0  1 Bpchange*Anes     0     0    -30;
Estimate 'C-H AT -30%' Anes 1 -1  0 Bpchange*Anes   -30    30      0;
Estimate 'C-P AT -30%' Anes 1  0 -1 Bpchange*Anes   -30     0     30;
Estimate 'H-P AT -30%' Anes 0  1 -1 Bpchange*Anes     0   -30     30;
```

Table 3 provides the SAS output from fitting this model. The sum of squares for BPCHANGE*ANES tests H_{01} and is significant. Thus we conclude that cycling frequency depends on blood pressure change. Table 3 contains the intercept and slope estimates for each anesthetic. Since the equal slopes model is not appropriate for these data, it is necessary to compare the anesthetics at various values of blood pressure change. We have chosen to compare the three anesthetics at BPCHANGE = −45.4% and −30.0%. The last 11 ESTIMATE statements are required to estimate the adjusted means and compare these means at each given value of blood pressure change. For example, "H-P at −30%" asks SAS to compare Halothane and Pentobarbital at a blood pressure change of −30%. The results of these tests are in Table 3 and a summary of results for both examples can be found in Table 4.

Table 3. *Example two. General linear models procedure.*

Dependent Variable: FREQ

Source	DF	Sum of Squares	Mean Square	F Value	Pr > F
Model	6	0.32077953	0.05346325	208.80	0.0001
Error	30	0.00768147	0.00025605		
Uncorrected Total	36	0.32846100			

R-Square	C.V.	Root MSE	FREQ Mean
0.342319	17.05819	0.0160015	0.0938056

Source	DF	Type III SS	Mean Square	F Value	Pr > F
ANES	3	0.02859050	0.00953017	37.22	0.0001
BPCHANGE* ANES	3	0.00219677	0.00073226	2.86	0.0534

Parameter	Estimate	T for HO: Parameter = 0	Pr > |T|	Std Error of Estimate
C Intercept	0.09020351	5.21	0.0001	0.01732883
H Intercept	0.07654260	3.16	0.0036	0.02420883
P Intercept	0.14407316	8.64	0.0001	0.01668434
C Slope	−0.00017283	−0.48	0.6353	0.00036073
H Slope	−0.00013625	−0.25	0.8063	0.00055069
P Slope	0.00100219	2.88	0.0073	0.00034810
C-H Slope	−0.00003658	−0.06	0.9561	0.00065832
C-P Slope	−0.00117502	−2.34	0.0259	0.00050130
H-P Slope	−0.00113844	−1.75	0.0908	0.00065148
C AT −45%	0.09805005	23.59	0.0001	0.00415624
H AT −45%	0.08272836	15.81	0.0001	0.00523128
P AT −45%	0.09857378	20.42	0.0001	0.00482756
C-H AT −45%	0.01532169	2.29	0.0290	0.00668136
H-P AT −45%	−0.01584542	−2.23	0.0337	0.00711840
C AT −30%	0.09538845	13.08	0.0001	0.00729092
H AT −30%	0.08063011	9.20	0.0001	0.00876223
P AT −30%	0.11400749	15.54	0.0001	0.00733767
C-H AT −30%	0.01475834	1.29	0.2053	0.01139887
C-P AT −30%	−0.01861904	−1.80	0.0819	0.01034403
H-P AT −30%	−0.03337738	−2.92	0.0066	0.01142882

CONCLUSION

Although our original hypothesis was that the occurrence of CBF oscillations was dependent on the type of anesthesia, as suggested previously [Clark Jr., Misrahy, and Fox, 1958; Hundley et al., 1988], this was not born out by the results. CBF cycling occurred under all the anesthetics used, Pentobarbital, α-Chloralose, and Halothane. For the variation of amplitude with blood pressure change, all the anesthetics demonstrated significant linear relationships. However, Pentobarbital showed a clear and significantly different linear relationship between blood pressure change and frequency of cycling from Halothane and α-Chloralose.

Table 4. *Summary statistics by type of anesthetic adjusting for blood pressure change using analysis of covariance.*

Variable	Anesthetic	n	Adjusted Mean ± SEM[†]	Slope ± SEE[‡]	Mean Blood Pressure Change (%)	Mean Square Error
Amplitude (%)	α-Chloralose	15	7.7 ± 0.7[a]	−0.114 ± 0.037	−45.4	6.73
	Halothane	10	6.2 ± 0.8	−0.114 ± 0.037		
	Pentobarbital	11	5.5 ± 0.8	−0.114 ± 0.037		
Frequency (Hz)	α-Chloralose	15	0.098 ± 0.004	−0.000173 ± 0.00036	−45.4	0.00256
	Halothane	10	0.083 ± 0.005[b]	−0.000136 ± 0.00055		
	Pentobarbital	11	0.099 ± 0.005	0.001002 ± 0.00035		
Frequency (Hz)	α-Chloralose	15	0.095 ± 0.007	−0.000173 ± 0.00036	−30.0	0.00256
	Halothane	10	0.081 ± 0.009	−0.000136 ± 0.00055		
	Pentobarbital	11	0.114 ± 0.007[c]	0.001002 ± 0.00035		

[†] SEM = Standard Error of Mean
[‡] SEE = Standard Error of Estimate
[a] Adjusted mean for Pentobarbital differs from that for α-Chloralose ($p = 0.04$).
[b] Adjusted mean for Halothane differs from that for α-Chloralose ($p = 0.03$) and Pentobarbital ($p = 0.03$).
[c] Adjusted mean for Pentobarbital differs from that for Halothane ($p = 0.007$).

REFERENCES

Chien, S (1985), *Cerebral blood flow and metabolism.* In: Principles of Neural Science, edited by Kandel, ER and Schwartz, JH. New York/Amsterdam:Elsevier, pp. 845–852.

Clark Jr., LC, Misrahy, G, and Fox, RP (1958), *Chronically implanted polarographic electrodes.* J. Appl. Physiol. 13: 85–91.

Frerichs, KU and Feuerstein, GZ (1990), *Laser-Doppler flowmetry. A review of its application for measuring cerebral and spinal cord blood flow.* [Review]. Mol. Chem. Neuropathol. 12:55–70.

Haberl, RL, Heizer, ML, Marmarou, A, and Ellis, EF (1989), *Laser-Doppler assessment of brain microcirculation: Effect of systemic alternations.* Amer. J. Physiol. 256:H1247–H1254.

Hudetz, AG, Roman, RJ, and Harder, DR (1992), *Spontaneous flow oscillations in the cerebral cortex during acute changes in mean arterial pressure.* J. Cereb. Blood Flow Metab. 12: 491–499.

Hundley, WG, Renaldo, GJ, Levasseur, JE, and Kontos, HA (1988), *Vasomotion in cerebral microcirculation of awake rabbits.* Amer. J. Physiol. 254: H67–H71.

Jöbsis, FF, Keizer, JH, LaManna, JC, and Rosenthal, M (1977), *Reflectance spectrophotometry of cytochrome aa_3 in vivo.* J. Appl. Physiol. 43:858–872.

Jones, SC, Williams, JL, Shea, M, Easley, KA, and Wei, D (1995), *Cortical cerebral blood flow cycling: Anesthesia and arterial blood pressure.* Amer. J. Physiol. 268: H569–H575.

Lindauer, U, Dirnagl, U, and Villringer, A (1993), *Characterization of the cerebral blood flow response to somatosensory stimulation with detailed description of the model and the influences of anesthetics.* Amer. J. Physiol. 264:H1223–H1228.

Mayevsky, A and Ziv, I (1991), *Oscillations of cortical oxidative metabolism and microcirculation in the ischemic brain*. Neurol. Res. 13:39–47.

Morita-Tsuzuki, Y, Bouskela, E, and Hardebo, JE (1992), *Vasomotion in the rat cerebral microcirculation recorded by laser-Doppler flowmetry*. Acta Physiol. Scand. 146: 431–439.

Stern, MD (1975), *In vivo evaluation of microcirculation by coherent light scattering*. Nature 254:56–58.

Vern, NS, Schuette, WH, Leheta, B, Juel, VC, and Radulovacki, M (1988), *Low-frequency oscillations of cortical oxidative metabolism in waking and sleep*. J. Cereb. Blood Flow Metab. 8:215–256.

Williams, JL, Shea, M, Wei, D, and Jones, SC (1992), *Cortical cerebral blood flow (CBF) cycling*. Soc. Neurosci. Abstr. 18: 405 (Abstract).

BIOGRAPHIES

Dr. Michael H. Kutner is the Chairman of the Department of Biostatistics and Epidemiology at The Cleveland Clinic Foundation. His research interests include regression diagnostics, nonlinear regression techniques, and statistical education. He is the coauthor of two popular statistical textbooks: *Applied Linear Regression Models* and *Applied Linear Statistical Models* published by Richard D. Irwin, Inc. (Neter, J, Kutner, MH, Nachtsheim, CJ, and Wasserman, W).

Kirk A. Easley is a Lead Biostatistician within the Department of Biostatistics and Epidemiology at The Cleveland Clinic Foundation. He is a statistician for a multicenter prospective natural history study funded by the National Heart, Blood and Lung Institute (NHLBI) that is designed to characterize the pulmonary and cardiovascular disorders that occur in children with vertically transmitted Human Immunodeficiency Virus (HIV) infection. His areas of interest include statistics in biomedical research and statistical problems in longitudinal data analysis.

Dr. Stephen C. Jones is the Director of the Cerebrovascular Research Laboratory and a member of the research staff in the Department of Biomedical Engineering at The Cleveland Clinic Foundation. His research interests are centered around cerebral blood flow and metabolism and include cerebral ischemia and cerebral blood flow control, the development and refinement of cerebral blood flow methodology and technique, and the investigation of therapeutic agents and diagnostic modalities for cerebral ischemia.

Dr. G. Rex Bryce is Chairman of the Department of Statistics at Brigham Young University. His research interests include linear statistical models, statistical collaboration, and statistical education. Dr. Bryce has pioneered efforts to develop strong academe/industry internships and partnerships.

CHAPTER 17

MODELING CIRCUIT BOARD YIELDS

Lorraine Denby, Karen Kafadar, and Tom Land

The manufacturing of products often involves a complicated process with many steps, the quality of which depends upon the complexity of the individual tasks. More complex components can, but need not, result in lower success rates in the final product. "Success" is measured differently for different products; it may be as simple as "unit turns on and off properly" or more quantitative such as "output power falls within the range 100 ± 0.05 watts."

The cost of poor quality is significant in a manufacturing plant: loss of a functioning unit that could have been sold for profit, lost employee time that produced the defective unit, diagnosing the problem, and correcting it if feasible, and materials from the unit that are no longer usable (scrapped). Thus, managers focus on ways of designing and building quality into the final product. If certain characteristics can be manufactured more successfully than others, it behooves the designer to incorporate such features wherever possible without sacrificing optimal performance. Statistical quality control usually involves the monitoring of product quality over time to ensure consistent performance of manufactured units. Our focus here is on quality one step before: to identify the characteristics of products which lead to higher probability of successful performance.

In this case study, we analyze the *yield* of printed circuit boards, i.e., the percent of boards in a production lot which function properly. Printed circuit boards are used in hundreds of electronic components, including computers, televisions, stereos, compact disk players, and control panels in automobiles and aircraft. Board failure means equipment failure, which is serious when the board controls the aircraft. Various characteristics of these boards can be identified, and board designers and manufacturers desire information on how the presence or absence of certain features can affect the yield. Features that lead to lower yields are candidates for either redesign or more careful analysis of the processes required to produce them.

INTRODUCTION

Product yields are important for cost, quality, and service in the printed circuit industry as well as other electronics manufacturing industries. Yield can be measured in a variety of ways. For the purpose of this study, *yield* is the ratio of number of boards shipped to number of boards started (*shipped + scrapped*).

In the printed circuits industry, the cost of poor quality may exceed 20% of sales. Scrap product is usually the largest component of that cost. In addition to process variations, product complexity will also influence yield. Complexity will vary considerably from one design to the next. Predicting yield as a function of product complexity becomes difficult because complexity can manifest itself as several dozen design variables, many of which are confounded with one another.

Predicting yields for new products can assist in price/cost decisions. Assessing what the yield "should be" for existing products can indicate which parts have abnormally low yields because of their complexity and which parts might have correctable problems related to some other factors. The parts with complexity-related yield problems need to be addressed with improved process capability. Those with other factors often need to be addressed through other means. The goals of this analysis are to identify which of the measured features on printed circuit boards have an effect on yield, either positively (raising yield) or negatively (reducing yield).

BACKGROUND INFORMATION

In the printed circuits manufacturing industry, almost all products are made to order. Each board design is unique, and hence the process that produces boards of a particular design has its own yield. When the manufacturer considers a new design for a customer, he must plan to price the product in accordance with yield expectations. If pricing assumes a 50% yield when in fact a 99% yield is achievable, a competing manufacturer may submit a lower bid and will likely receive the contract. Conversely, if the manufacturer prices the product expecting to achieve 99% yield but achieves only 50%, the financial losses will be severe. Most manufacturers "guess" at what the yield will be for new products based on opinion and experience. These guesses often err by 5% to 10% and occasionally miss by even greater margins.

Each design could be characterized with many variables. Some are known early and are easy to measure. Others are not measured until after a price quotation has been delivered. Still other variables are not easily measured or may be rarely measured at all. Most of these variables are confounded with one another. More complex designs often have more layers, narrower lines, more holes, and smaller holes than do less complex designs. Each customer uses different CAD software and presents a unique set of finished product requirements; consequently, customer identity also contributes to product variability.

Process variation complicates matters further. One lot of the same design might be processed at 100% yield while another lot a few days later might be scrapped completely (0% yield). When comparing results from two different products, the analyst cannot be sure if the differences are a result of product variation (design) or process variation.

The goal of this project is to develop a simple but accurate prediction of yield for each new design. This prediction should be based on variables/attributes that are easily measured and are known early in the process. Inaccurate predictions are no better than the current "guessing" system, and complicated algorithms and predictions based on variables that have not been measured cannot be used.

QUESTIONS OF INTEREST

The engineers are particularly interested in being able to predict the yields of printed circuit boards given information on various characteristics of the boards type. Some of these characteristics are assumed to be associated with board complexity, such as number of layers, minimum drill hole size, the presence of gold-plated fingers; one might suppose that more complex boards have lower yields. To some extent, the engineers might want to know which variables are most influential in predicting yield. These predictions allow the engineers to plan their production needs and schedule the manufacturing of boards accordingly and price the boards appropriately.

DATA

Measured characteristics and performance yields for 96 types of printed circuit boards were collected over a six-month period (April–September). Although weekly data are available (number of boards *shipped* and *scrapped*), from which weekly performance yields can be calculated [*yield = #shipped/(#shipped + #scrapped)*], we confine ourselves to only the overall yield for the entire 6-month (26-week) period. The characteristics of the board types are as follows:

1. *testpts* = the number of endpoints (e.g., connections) that the board (part number) must pass in order for the board to be shipped. This may be the best estimate of the total functionality of the part and is probably confounded with most of the other variables. An individual with cost/price responsibility for new part numbers would usually know all of the other variables in this list before knowing this variable. Yield is expected to decrease with an increased number of test endpoints.

2. *subp* = the number of subpanels per board. Given a constant panel size, the number of subpanels per panel is inversely related to the area of product that has to be good in order for the part to ship. Yield is expected to increase with the number of subpanels per panel. The area of the board is approximately 340/*subp*.

3. *small* = number of small (25 mm or less) holes in a subpanel. Yield is expected to decrease with an increased number of small holes. *small/subp* gives the average number of small holes per subpanel.

4. *large* = number of large (25 mm or more) holes in a subpanel. Yield is expected to decrease with an increased number of large holes. *large/subp* gives the average number of large holes per subpanel.

5. *lyrs* = number of layers in the part number. Generally, multilayer boards are more complicated to manufacture.

6. *ess* = binary variable: 1 indicates that the part number uses this type of solder mask coating; 0 indicates that it uses one of the other three possible types. Generally, part numbers using the ESS solder mask coating have had lower yields than those using one of the other types.

7. *gold* = binary variable: 1 indicates that the part number has gold fingers, 0 indicates that it does not. The presence of gold fingers presents additional failure modes to which nongold boards are not subject. Yield is expected to be better on nongold products than it is on gold products.

8. *min.drill* = the smallest size hole that is drilled into a given part number. Yield is expected to decrease as hole size decreases, since smaller drill hole sizes are more complicated to manufacture.

9. *smdra* = binary variable: 1 indicates board types with a two mil solder mask registration allowance. Such boards are processed on superior equipment. They have the disadvantage of tighter tolerances but the advantage of the superior equipment. The relationship between yield and this variable could be either favorable or unfavorable.

10. *ci.tol* = binary variable: 1 indicates that the board type has a tight tolerance for controlled impedance, 0 indicates normal tolerance. In addition to the standard electrical test, some parts have a controlled impedance requirement that subjects them to further testing and additional failure modes to which normal product is not subject. Yield is expected to be lower on parts with a controlled impedance requirement.

11. *oem* = binary variable: 1 indicates that the board type is manufactured for an outside equipment manufacturer, 0 indicates internal use. Most of the factory's products are sold to other divisions of the company. Some are sold on the market to "OEM" customers. OEM parts typically have somewhat different endpoint requirements and less robust designs which produce lower yields.

12. *int.lw* = the circuit width on internal layers of the board. This variable is not applicable to two layer boards as there are no internal layers. This variable has four levels: 5, 6, > 6 (greater than 6), and "NA" (not applicable for boards having only two layers; see variable *lyrs2*). Thus, only two of the three contrasts associated with this variable will be meaningful and should be included in the analysis. Yield is expected to decrease as the circuit width becomes smaller.

13. *lyrs2* = binary variable indicating that the board has only two layers. Notice that we need this variable because *int.lw* is not meaningful for boards having only two layers, since there are no internal lines on such boards.

14. *ext.lw* = the circuit width on external layers of the board. Yield is expected to decrease as the circuit width becomes smaller.

15. *ship.t* = total number of boards shipped during the six-month period.

16. *scrap.t* = total number of boards scrapped during the six-month period.

17. *yield.t* = total yield, *ship.t/(ship.t + scrap.t)*, over the six-month period.

A subset of the data follows.

	C1	C2	C3	C4	C5
testpts	4299	4010	2038	5704	1568
lyrs	6	6	6	6	6
ess	0	0	0	0	0
subp	1	3	3	4	4
gold	1	0	1	0	1
min.drill	25	13	22	13	25
smdra	0	0	0	0	0
ci.tol	0	0	0	0	0
oem	0	0	0	0	0
small	4069	15601	1830	19243	2611
large	139	164	5906	3104	5676
ship.t	162850	133532	114629	346095	224070
scrap.t	62150	51458	38194	117055	81330
yield.t	0.7238	0.7218	0.7501	0.7473	0.7337

QUESTIONS FOR ANALYSIS

1. What is the distribution of layers (*lyrs*) in our data set (i.e., how many boards have two layers, three layers, etc.)? Likewise, how many boards
   - use/do not use the newest solder mask (*ess*);
   - require/do not require gold-plated fingers (*gold*);
   - are for an outside/inside equipment manufacturer (*oem*);
   - have a tight/normal tolerance for controlled impedance (*ci.tol*);
   - have tight/normal solder mask registration allowance (*smdra*).
2. Construct either a histogram or a stem-and-leaf diagram for the variables *testpts, large, small, subp, min.drill*. What do you conclude about the distributions of these variables in our data set?
3. When the response variable of interest is a proportion, or yield, say *p*, we often find it useful to consider transforming the proportion via the transformation: $logit(p) \equiv log\ [p/(1-p)]$. In medical studies, this transformed variable is called the logarithm of the odds ratio (the odds of survival versus not surviving); it is also the "natural parameter" when the binomial family is expressed in exponential family form. Construct boxplots for $logit(yield.t)$ versus (a) number of layers (*lyrs*), (b) outside equipment manufacturer (*oem*), (c) gold fingers (*gold*), (d) solder mask registration allowance (*smdra*), (e) use of newest solder mask (*ess*), (f) internal line width (*int.lw*), (g) external line width (*ext.lw*), and (h) controlled impedence (*ci.tol*). Because of the distribution of these variables in the data set, it is helpful to construct boxplots having variable widths (i.e., width proportional to number of observations in the box; see [McGill, R., Tukey, J.W., Larsen, W.A., 1978]). Which variables appear to have an effect on the yield, and in what direction?
4. Some of the variables may interact with respect to yield. Calculate weighted yields for the four categories *oem* (0 or 1) × *gold* (0 or 1). This can be done most effectively by defining a variable *N.t = ship.t + scrap.t* for the total number of boards for each *oem* × *gold* cell. For example, the yield when *gold* = 0 *and oem* = 1 can be calculated as

$$p_{01} = \sum_j N.t(j) \times yield.t(j) / n_{01},$$

where $n_{01} = \sum_j N.t(j)$ and the summation $\sum_j$ is over all board types having *gold* = 0 and *oem* = 1.

Note that p_{01} is equivalent to the number of boards shipped over the total number manufactured (*ship.t* + *scrap.t*), amongst all those for which *gold* = 0 and *oem* = 1. Amongst all boards manufactured for internal use (*oem* = 0), do boards with gold fingers have higher or lower yields than those without gold fingers?

Recall from one of your early statistics classes (or see [Snedecor, G. and Cochran, W.G., 1967, section 8.7, p. 210]) that an approximate standard error of the estimate of a binomial proportion, *yield.t(j)*, can be calculated as $[y_j(1-y_j)/N.t(j)]^{1/2}$, where $y_j \equiv yield.t(j)$. Using this formula, calculate the standard error of the weighted mean yield p_{01} given above, as well as those for the other three categories. (Because the $N.t(j)$'s are so large, these standard errors, while correct, are of limited usefulness in assessing the significance of the differences in yields in different categories.) Is there an interaction between *gold* and *oem*? Answer the same question for the variables *ess* and *gold and* for *lyrs* and *gold*.

5. The engineers would prefer not to require the variable *testpts* in a prediction model, because the value of this variable is known only long after the board is designed. However, the number of holes on the board per unit area is often a good indicator of how many points will need to be tested. Recall that the area is inversely proportional to *subp*. Carry out a linear regression of the variable *testpts* on the two variables x_1 = *small/subp* and x_2 = *large/subp*. Plot *testpts* as a function of $x_1 + x_2$. Do you notice any potential outliers? Now construct a normal quantile–quantile (qq) plot of the residuals from your model. Run another regression model without the outlier(s). How much variation is explained by your model? Can you conclude the adequacy of a linear combination of x_1 and x_2 in place of *testpts* for a linear prediction model of yield?

Notes for Questions 6–8

An important consideration in fitting a linear regression model is the appropriateness of the assumptions. Classical linear regression assumes that the response y_i is normally distributed with mean $x_i'\beta$ and variance σ^2, independent of i, for $i = 1 \ldots n$ (=96 here), where $x_i' = (x_{01}, x_{1i}, K, x_{ki})$ are the measured characteristics of board type i and β is a $(k+1)$-dimensional vector of unknown coefficients to be estimated (e.g., via least squares). Clearly these assumptions do not hold for our case, since our y_i is a yield. There are two approaches to handling this situation:

A. Use a generalized linear model routine. This allows you to state explicitly that the distribution of the random variable $N.t(i) \cdot p_i$, $i = 1, K, 96$, is binomial with parameters $N.t(i)$ and ρ_i, where $\rho_i = x_i'\beta$ is the linear combination of our explanatory variables that we use to predict y_i. We aim to estimate β for good predictions based on our choice of variables.

B. Some packages may not have a generalized linear model routine. We can effect the same procedure as follows. Recall that a proportion, y_i, has a distribution that is asymptotically normal with mean ρ_i and variance $\rho_i(1-\rho_i)/N.t(i)$, especially if $N.t(i)$ is large. (In our case, all of the $N.t(i)'s$ are well over 1000, so our approximation to normality is quite safe.) We can now use an ordinary linear regression routine but insist that the routine recognizes the differences in the variances of our y_i, which we approximate by $[y_i(1-y_i)/N.t(i)]$. The optimal weights for a linear regression whose observations have unequal variances are inversely proportional to the variances; hence, $w_i^{(0)} = N.t(i)/[y_i(1-y_i)]$. Let us call the predicted y_i from such a model $\hat{y}_i^{(1)}$.

Obviously, we would like to use the true ρ_i instead of the sample observed y_i, so we can improve our estimate of β by updating our weights with $w_i^{(1)} = N.t(i)/[\hat{y}_i^{(1)}(1-\hat{y}_i^{(1)})]$. The predicted y_i from this second model is $\hat{y}_i^{(2)}$, and of course we can continue by setting $w_i^{(2)} = N.t(i)/[\hat{y}_i^{(2)}(1-\hat{y}_i^{(2)})]$ until there is no further change in our fitted model coefficients. (A generalized linear model routine effectively executes these iterations.)

6. Carry out a linear regression according to prescription (B) above with the following variables:

y_i = dependent variable, *yield* of board type i, $x_1 = small/subp$, $x_2 = large/subp$,

$x_3 = lyrs$, $x_4 = gold$, $x_5 = ci.tol$, $x_6 = lyrs$, $x_7 = int.lw$, $x_8 = lyrs2$, $x_9 = ext.lw$,

$x_{10} = smdra$, $x_{11} = oem$, $x_{12} = oem \times gold$.

Notes for Question 7

You will need 3–4 iterations. You will also need to specify that only two of the three contrasts associated with the variable *int.lw* are meaningful: one for internal line widths of 5 and one for internal line widths of 6. The coefficient for the factor corresponding to > 6 will be the negative of the sum of the coefficients for these two variables, and the third contrast must be left out of the analysis. You can accomplish this by specifying formally two coded variables, say (z_1, z_2), taking the values (z_1, z_2) = (1,0) if *int.lw* = 5, (0,1) if *int.lw* = 6, (−1,−1) if *int.lw* >6 , and (0,0) if *int.lw* = NA, i.e., missing. In S or S-PLUS, set *options (contrasts = 'contr.sum')* and include as a variable *con.lw ← C(int.lw, contrasts(int.lw), 2)* in your model statement (call for *lm*). Which variables are highly significant in predicting *yield*? What do the coefficients indicate about their effects on *yield*?

How much variation is explained by your model? If any variables appear to be nonsignificant, try refitting your model without them. State clearly your final model and the importance of the variables included in it.

7. Construct the analysis of variance table corresponding to the variables in your model. (If you coded the contrasts for *int.lw* directly, remember to combine the sum of squares for these two variables for the total sum of squares for the variable *int.lw*.) Most statistical packages will compute the *p*-values hierarchically; i.e., the additional reduction in the sum of squares given the preceding terms in the model.

8. Plot the residuals from your model versus the fitted values. Are there any patterns in your plot? Also construct boxplots of the residuals for the variables of greatest significance (p-value less than 0.05). Are the residuals uniformly distributed over different levels of the variables?

INSTRUCTIONS FOR PRESENTATION OF RESULTS

Prepare a one-page executive summary for the manufacturing manager of this project, stating your conclusions from this analysis. Remember that the manager is not likely to be a statistician, so you will need to explain your results in manufacturing language.

REFERENCES

McCullagh, P. and Nelder, J.A. (1983), *Generalized Linear Models*, Chapman & Hall, London.

McGill, R., Tukey, J.W., and Larsen, W.A. (1978), Variations of box plots, *American Statistician* 32, pp. 12–16.

Neter, J., Kutner, M., Nachsheim, C.J., and Wasserman, W. (1994), *Applied Linear Statistical Models, Fourth Edition*, Irwin, Chicago.

Snedecor, G. and Cochran, W.G. (1967), *Statistical Methods, Sixth Edition*, Iowa State University Press, Ames, Iowa.

BIOGRAPHIES

Lorraine Denby is a member of the Technical Staff at Bell Labs—Lucent Technologies. She received her Ph.D. in statistics from the University of Michigan. She is a Fellow of the American Statistical Association and is an elected member of the International Statistics Institute and has served as a member of the American Statistical Association Board of Directors.

Karen Kafadar is Professor of Mathematics at the University of Colorado-Denver. She received her B.S. and M.S. from Stanford and her Ph.D. in Statistics from Princeton. Prior to her position at the University, she spent three years in the Statistical Engineering Division at the National Bureau of Standards (now National Institute of Standards and Technology), seven years in the research and development laboratory at Hewlett Packard Company in Palo Alto, and three years in the Biometry Branch at the National Cancer Institute. She has been awarded the Wilcoxon prize and the Outstanding Statistical Applications award in 1995 and is a Fellow of the American Statistical Association.

Tom Land is a Member of Technical Staff at Lucent Technologies Atlanta Works. At the time of this writing, he worked at AT&T's Richmond Works. He has eleven years experience as a Quality Engineer. Most of that effort has been yield improvement in Printed Circuits manufacturing. He has a Bachelor's Degree in Chemical Engineering from Virginia Tech, an MBA from the University of Richmond, and is currently pursuing a Master's Degree in Industrial Engineering from Purdue University.

CHAPTER 18

EXPERIMENTAL DESIGN FOR PROCESS SETTINGS IN AIRCRAFT MANUFACTURING

Roger M. Sauter and Russell V. Lenth

This case study is about designing and analyzing experiments that are relevant to hole-drilling operations in aircraft. When a change was made to a new lubricant, it was necessary to do some experimentation to learn how much of this lubricant should be used and how it interacts with other process factors such as drill speed. Several factors are involved in our experiment, and there are physical and time constraints as well, necessitating an incomplete-block experiment where only a subset of the factor combinations are used on any one test coupon. The reader is guided through the design and analysis, considering some related practical issues along the way.

INTRODUCTION

The goal of this study is to design and analyze an experiment that will help improve a manufacturing process—in this case, the assembly of aircraft.

The skin of the fuselage (i.e., the body of the plane—see Figure 1) and wings are made of pieces of metal, overlapped and fastened together with rivets. Thus, the process involves drilling a very large number of holes. These must be positioned accurately, and the quality of the holes themselves is important.

The experiment in this study was motivated by a change in lubricant. A certain amount of lubrication reduces friction, prevents excessive heat, and improves hole quality; however, too much lubricant can be problematic because the drill bit may not get enough "bite." In the past, chlorofluorocarbons (CFCs) had been used as a lubricant. These can no longer be used due to environmental concerns, necessitating the use of a new class of lubricants. Changing the lubricant can have a far-reaching impact on the entire process; hence, experimental methods are used to study the key control variables—including the amount of lubricant—in the hole-drilling process.

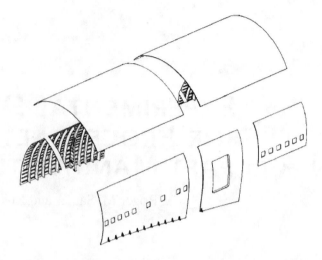

Fig. 1. *Panels in part of the fuselage.*

The response variables (outputs) of interest are the important quality characteristics of holes. These include the hole diameter and the surface finish inside the hole. These are described in more detail below.

BACKGROUND

By definition, experimentation entails trying different settings to see what works and what doesn't. The experiment will not provide much useful information unless it produces a range of hole qualities: we have to try to produce some bad holes along with some good ones. For this reason as well as cost, we do not want to run such experiments on actual aircraft assemblies. Instead, the experiments are run off line, using small pieces of material called test coupons.

Measuring Hole Quality

The customer for this experiment is most interested in how various factors affect the diameter and surface finish of the hole. Accordingly, the surface finish and diameter are the primary measures of hole quality and serve as the response variables in our experiment.

Surface finish is measured using a probe that can detect the minute peaks and valleys of the surface inside the hole along the path of the probe. The roughness, denoted R_a, of the hole can be quantified using the total area from the average depth. Specifically, R_a is defined as the total area of the absolute deviations of the probe from its average depth—the shaded area in Figure 2. The acceptable magnitude of these peaks and valleys varies depending on the engineering requirements for a given installation.

Fig. 2. *Illustration of the definition of R_a.*

Hole diameter is measured using an instrument that is pushed into the hole. It measures the hole diameter at different depths and in different directions; these numbers are averaged together into one measurement of hole diameter. Instead of reporting the raw measurement, the diameter of the drill bit used to drill the hole is subtracted from the measured hole diameter, yielding the "excess diameter" values in the data set.

Factors

The experiment contains four factors: the rotational speed of the drill (RPM, rotations per minute, at three levels), the feed rate of the drill (IPR, for inches per revolution, at three levels), the temperature of the air jet (AirTemp, in degrees Fahrenheit, at two levels), and the volume of lubricant (LubeVol, in milliliters, at two levels). There are thus a total of $3 \times 3 \times 2 \times 2 = 36$ factor combinations.

Six test coupons (Coupon) are used in the experiment. It is not possible to run all 36 factor combinations on 1 coupon, so only 6 factor combinations are run on each coupon. (This makes it an incomplete-block experiment.) The factor combinations are assigned so that a balance of levels of each factor is represented on each coupon. Refer to the data file described below to see exactly how these assignments were made.

QUESTION OF INTEREST

Our goal is to find the settings of speed, feed rate, air temperature, and lubrication that yield the best hole quality. (What is meant by "best" is discussed later.) This small experiment is unlikely to answer all of our questions, so another goal is to recommend a new experiment that will help shed further light in the quest to improve the hole-drilling process.

DATA

Name of data file: Case18.txt

Variable name	Description
RPM	Drill speed in revolutions per minute (3 levels: 6,000, 10,000, and 15,000)
IPR	Feed rate in inches per revolution (3 levels: .006, .009, .012)
AirTemp	Temperature of the air jet in degrees Fahrenheit (2 levels: 30 and 50)
LubeVol	Amount of lubricant, in milliliters (2 levels: .01 and .02)
EntRa	Average entrance surface finish (R_a) for holes 1, 5, 10, 15, 20
ExitRa	Average exit surface finish (R_a) for holes 1, 5, 10, 15, 20
EntLnRa	Average entrance ln surface finish (ln R_a) for holes 1, 5, 10, 15, 20
ExitLnRa	Average exit surface finish (ln R_a) for holes 1, 5, 10, 15, 20
XS_Diam	Average hole diameter minus drill diameter for holes 1, 5, 10, 15, 20

Coupon	RPM	IPR	Air Temp	Lube Vol	Ent Ra	Exit Ra	Ent LnRa	Exit LnRa	XS_ Diam
1	6000	0.006	30	0.02	21.6	21.4	3.040	3.027	1.21
1	15000	0.009	50	0.02	23.4	21.4	3.028	3.023	1.32
1	10000	0.012	30	0.02	23.6	23.4	3.010	3.125	1.33

The experiment involved drilling a sequence of 20 holes at each factor combination. Only the 1st, 5th, 10th, 15th, and 20th hole in each sequence was actually measured; hence, the complete data set consists of measurements on $5 \times 36 = 180$ holes. Picking these hole numbers provides an opportunity to examine heat build up and its effect on hole quality over each group of the 20 holes (in a different analysis than this case study). Summary statistics were computed for each set of 5 measured holes. Specifically, the following summary statistics are in the data set: the mean of the surface finish (R_a) at the entrance and the exit of the hole; the mean of the natural logarithm of R_a at the entrance and exit; and the mean excess diameter. All analyses in this case study are based on these summary values.

GUIDED ANALYSIS

Design Concepts

Since one of the interesting features of this study is the experimental design itself, we ask the student to begin by answering some practical questions about the measurements and the design.
1. Suppose that you have measured the excess diameter of two different holes.
    A. How would you know which hole is of higher quality? What if the measurement were the surface finish (R_a)? Do you think that excess diameter and surface finish will be correlated?
    B. If there is substantial variation in one of these measurements (under the same operating conditions), is that good, bad, or unimportant?
    C. Explain why excess diameter is a better measure of quality than just the diameter of the hole.
    D. Can you think of other useful ways of measuring hole quality?
2. Suppose that the experiment involves six coupons of the same type and thickness of metal. In selecting coupons, should we get six coupons from the same batch of incoming material, or should we get the coupons from different batches? Discuss the advantages and disadvantages of each.
3. Consider a different (simpler) experiment that involves six coupons from different batches of material. We want to test the three different drill speeds—6000, 10000, and 15000 RPM. We can drill 60 holes in each coupon. Discuss which is better:
    A. Randomly assign two coupons to 6000 RPM, two others to 10000 RPM, and the remaining two to 15000 RPM. Drill all 60 holes in each coupon at its assigned speed.
    B. In each coupon, drill a row of 20 holes at 6000 RPM, 20 holes at 10000 RPM, and 20 holes at 15000 RPM. A random mechanism (separate for each coupon) is used to decide which speed comes first, second, and third.

4. The plan in 3(b) calls for all 20 holes at each speed to be drilled in sequence on each coupon. Another possibility—call it plan 3(c)—allows the speed to change between individual holes so that for each coupon, some random subset of 20 of the 60 holes are drilled at 6000 RPM, a different 20 are drilled at 10000 RPM, and the rest are drilled at 15000 RPM. The 60 holes are then drilled in random sequence. Discuss the relative merits of this randomization compared with the one in 3(b), taking both the statistical and the practical issues into account.

5. The experimental procedure used in this case study was as follows:
   - We do all test runs on a coupon before proceeding to the next coupon.
   - The order in which coupons are tested is randomized.
   - For each factor combination, a row of 20 holes is drilled in sequence on the coupon.
   - It was not possible to use the same drill bit for all of the holes; therefore, six drill bits were used on each coupon and the bits were randomly assigned to the six factor combinations. We consider the drill bits to be nearly identical in quality, but there is slight natural variation in their diameters. The diameter of each drill bit was measured and used as an adjustment to the measurements of hole diameter.
   - Within each coupon, the order of testing its six assigned factor combinations is randomized. (The order of testing also defines the physical position of the row of holes on the test coupon.)
     What has been lost by testing one coupon at a time, rather than mixing them up?

6. By reducing the data to summary measures, we now have only one observation per factor combination. Is there any hope of obtaining valid statistical results without any replications?

Data Analysis

1. Construct an analysis of variance table for ExitRa, using a model with a main effect of Coupon, plus all main effects and two-way interactions of RPM, IPR, AirTemp, and LubeVol. Construct and interpret diagnostic plots of the residuals. Do the same using ExitLnRa as the response. (Note: Because this is an incomplete-block experiment, computer procedures designed solely for balanced analysis of variance will balk at these data. You will need to use a "general linear models" procedure, or do regression on dummy variables.

2. Based on the above analyses, decide whether you want to continue to work with ExitRa or ExitLnRa as the response variable. Construct interaction plots for the mean response at combinations of each pair of factors RPM, IPR, AirTemp, and LubeVol. Based only on looking at these plots,
   A. What factor combination appears to yield the best overall quality?
   B. What factors interact the most strongly?

3. Complete the construction of interaction plots, analysis of variance tables, and diagnostic plots for the response variables EntRa (or EntLnRa) and XS_Diam. Determine the best combinations (and the close runners-up) for each response variable. Are there any serious contradictions among the best combinations, depending on the response variable?

Further Reading

Most standard design texts—for example, [Montgomery, 1997, pp. 208–219]—discuss incomplete-block designs in one factor. Somewhat fewer have material on multifactor designs in incomplete blocks. Material on such designs having all factors at two or three levels can be found in [Hicks, 1993, sections 12.3 and 13.5] and [Montgomery, 1997, Chap. 8 and section 10-2]. An example of an experimental design closely related to ours (a $3 \times 3 \times 2$ design in 3 blocks of size 6) is discussed in [Cochran and Cox, 1957, pp. 203–212]. The design in this study incorporates restrictions on randomization—a common practice. [Lorenzen and Anderson, 1993] is a good reference for this topic.

PRESENTATION OF RESULTS

Based on the analysis above, write a short report and prepare a presentation directed to the engineer in charge that summarizes your conclusions and recommendations. The report should be no more than three pages in length and should contain only pertinent graphs and tables to aid in the engineer's understanding. Organize your own work so that if the engineer were to come back in six months with a question, you would know what you did and did not do, and what methods you used.

REFERENCES

Cochran, William G. and Cox, Gertrude M. (1957), *Experimental Designs*, 2nd ed., New York: John Wiley and Sons.

Hicks, Charles R. (1993), *Fundamental Concepts in the Design of Experiments*, 4th ed., New York: Saunders College Publishing.

Lorenzen, Thomas J. and Anderson, Virgil L. (1993), *Design of Experiments: A No-Name Approach*, New York: Marcel Dekker.

Montgomery, Douglas C. (1997), *Design and Analysis of Experiments*, 4th ed., New York: John Wiley and Sons.

BIOGRAPHIES

Roger Sauter serves as a statistical consultant for Boeing Commercial Airplane Group. His areas of interest include experimental design and interval estimation. He received his Ph.D. in statistics from Oregon State University and his MS in statistics from Kansas State University.

Russell Lenth is an Associate Professor of Statistics and the director of the Program in Quality Management and Productivity at the University of Iowa. He earned his MS and Ph.D. degrees in Mathematics and Statistics at the University of New Mexico, Albuquerque. His research interests include experimental design, statistical computing, and power analysis.

CHAPTER 19

AN EVALUATION OF PROCESS CAPABILITY FOR A FUEL INJECTOR PROCESS USING MONTE CARLO SIMULATION

Carl Lee and Gus A. D. Matzo

Capability indices are widely used in industry for investigating how capable a manufacturing process is for producing products that confirm the engineer's specification limits for essential quality characteristics. Companies use them to demonstrate the quality of their products. Vendees use them to decide their business relationship with the manufacturer. One important underlying assumption for capability analysis is that the quality characteristic should follow a normal distribution. Unfortunately, many quality characteristics do not meet this assumption. For example, the leakage from a fuel injector follows a very right-skewed distribution. Most of the leakages are less than one ml, with a few cases over one ml and some rare cases over three ml. It is important to understand how well these indices perform if the underlying distribution is skewed. This case study is initiated from the concerns of using these indices for reporting the capability of a fuel injector process in an engine manufacturing plant.

INTRODUCTION

The fuel injection system of an automobile meters fuel into the incoming air stream, in accordance with engine speed and load, and distributes this mixture uniformly to the individual engine cylinders. Each cylinder is connected with an injector. When an engine is turned on, fuel injectors inject fuel into individual cylinder along with the incoming air to form a uniform mixture for ignition. When the engine is turned off, the injector should stop injecting fuel immediately. However, during the first few seconds after turning off the engine, a tiny amount of fuel may leak out from the injector into the engine cylinder. This is the injector leakage. Such leakage is undesirable from an emissions standpoint.

The amount of leakage is usually less than one ml with rare occasions over three ml. The distribution of leakage is skewed to the right. The leakage is monitored using statistical process control (SPC) methods, and the capability indices are used to measure if the amount of leakage meets the engineer's specification limits. Due to the fact that the distribution of leakage is highly skewed, there has been a concern of the validity of using the capability indices that are developed for variables that follow normal curves. In fact, this is not an isolated case. Quality characteristics in industrial environments often do not follow normal distributions. This case study is conducted for two main purposes:
1. To explore the quantitative properties of the fuel injector's leakage using a set of leakage data collected at a fuel injector flow laboratory at FORD.
2. To investigate the performance of capability indices when the underlying distribution of a quality characteristic is highly skewed. This purpose is not limited to the leakage problem. It has a more general goal in mind to study the performance of capability indices under different degrees of skewness and several other factors that may also have effects.

BACKGROUND INFORMATION

The automotive industry is complex. There are thousands of components in an automobile. Each component must be at a high quality level in order to have a quality automobile. Many technologies developed by the automotive industry are also the foundation of many other manufacturing industries. Since the 1980s, the automotive companies have been undergoing a very deep restructuring to attain high quality automobiles for the highly competitive international market. One of the major tools in pursuing quality improvement is statistical methodology. Among the essential tools are graphical tools such as the histogram, probability plot, box plot, pareto diagram, cause-effect diagram, statistical process control charts, capability analysis, and more advanced statistical methods such as design of experiments, analysis of variance, regression techniques, simulation techniques, etc.

Most automobile engines are four strokes where the engine operates on what is known as the four-stroke cycle (see Figure 1). Each cylinder requires four strokes of its piston to complete the sequence of events which produces one power stroke. These four strokes are (1) intake stroke, (2) compression stroke, (3) power stroke, and (4) exhaust stroke. The intake stroke draws a fresh mixture of fuel and air into the cylinder. The compression stroke compresses the mixture to the point where cylinder pressure rises rapidly. At the power stroke, as the temperature and pressure rise, combusted gases push the piston down and force the crankshaft to rotate. At the exhaust stroke, the burned gases exit the cylinder, and another cycle begins.

The fuel injection system controls the amount of fuel going into the air mixture at the intake stroke stage. Fuel is injected through individual injectors from a low-pressure fuel supply system into each intake port. The ratio of mass flow of air to mass flow of fuel must be held approximately constant at about 14.6 to ensure reliable combustion. Figure 2 shows the detailed components of a fuel injector. When the engine is shut off, the injector should stop. Ideally, there should be not any fuel leakage from injectors. However, the physical design is so complex that it may be impossible to prevent some leakage.

Although the injector leakage reduces rapidly with engine usage, strict emission regulations require that the leakage be low over the entire vehicle life, starting at virtually zero miles. High leakage equates to potentially higher undesirable tailpipe emissions.

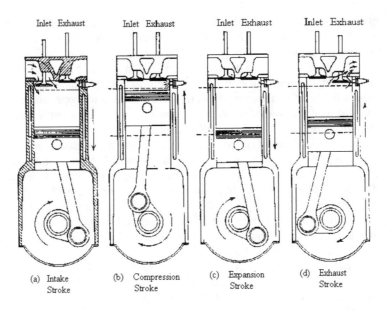

Fig. 1. *The four-stroke cycle.*

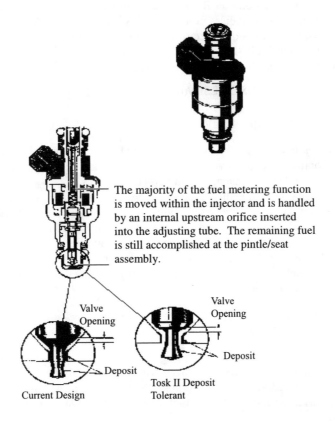

The majority of the fuel metering function is moved within the injector and is handled by an internal upstream orifice inserted into the adjusting tube. The remaining fuel is still accomplished at the pintle/seat assembly.

Fig. 2. *The cross section of a fuel injector.*

SPC charts are used to monitor the special causes of a manufacturing process. If any out-of-control condition shows on the charts, managers will look for some special causes associated with these out-of-control problems. The capability analysis studies the ability of the process to produce products within the specification limits. Specification limits are designed to meet engineering and emission regulation consideration. Capability analysis evaluates the accumulated variation of the entire process when it is operated under normal circumstance. It provides information about the noise level of the system when there is no out-of-control condition indicated on the SPC charts.

The sampling procedure of collecting leakage data for process monitoring and capability analysis is the following. Ten injectors are sampled at a randomly selected time period as a subgroup. They are brought into a laboratory and automatically connected to a manifold. The manifold is put under controlled liquid pressure, and held. The injectors are operated for a few seconds, as they would in the engine. Leaking fluid is measured at the nozzle tip of each injector. The entire test is controlled and measured by a computer. The leakage data are used for monitoring the injector production line and for capability analysis.

The fuel injector leakage data are used in this case study because they represent a general class of distribution types frequently encountered in SPC applications, i.e., unilateral (one-sided) and positively (right) skewed. In the case of fuel injectors, leakage cannot be negative.

Commonly used SPC control charts for variable data are $\overline{X}$ - and R-charts. An $\overline{X}$ - chart monitors the process mean and an R-chart monitors the process variation along the time domain. Commonly used indices for capability analysis are C_p, C_{pk} for quality characteristics requiring two-sided specification limits, C_{pl} for characteristics requiring only the lower specification limit (LSL), and C_{pu} for characteristics requiring only the upper specification limit (USL). These indices are defined in the following:

$$C_p = (USL - LSL)/(6\sigma), \quad C_{pu} = (USL - \mu)/(3\sigma),$$

$$C_{pl} = (\mu - LSL)/(3\sigma), \quad C_{pk} = \min \{C_{pl}, C_{pu}\},$$

where the notation μ and σ are the process mean and process standard deviation. Virtually every quality control textbook covers the SPC control charts and capability analysis (e.g., [Montgomery, 1996; Devor, Chang, and Sutherland, 1992]).

Both SPC charts and capability indices are designed for characteristics that follow a normal distribution. However, this assumption may not be valid in real world applications. One example is the amount of leakage of a fuel injector, which is highly skewed. Therefore, it is important to investigate "how well these indices and/or SPC charts perform when the underlying distribution is highly positively skewed." Gunter had a series of four articles on the use and abuse of C_{pk}. The methodology presented in this study is not limited to the injector leakage problem.

QUESTIONS OF INTEREST

There are two purposes for this case study. One is to explore the quantitative properties and the distribution of the fuel injector leakage; the other is to investigate the performance of some commonly used capability indices when the underlying distribution is highly positively skewed. To address the first purpose, the following questions are asked:

1. What is the average and the median fuel injector leakage? What is the leakage standard deviation?

2.   Does the distribution of the leakage follow a normal curve? If not, how skewed is it? What type of distribution is it?

These questions are investigated using the leakage data of 94 randomly tested fuel injectors.

For the fuel injector process, the smaller the leakage, the better the injector. The best would be zero leakage. Therefore, it is only necessary to have the upper specification limit, and C_{pu} is usually used for the capability analysis. If the quality characteristic has a nominal value, then two-sided specification limits are necessary. Under this situation, C_p and C_{pk} are commonly used. If the quality characteristics are "the-higher-the-better," one needs only the lower specification limit, and C_{pl} is commonly used. This case study focuses on the performance of C_p, C_{pu}, and C_{pk}, since C_{pl} is not used only for quality characteristics that are the "the-smaller-the-better."

In a typical process capability analysis for a manufacturing production line, the observations collected for SPC control charts are also used for capability analysis. The procedure for constructing SPC control charts involves the determination of subgroup size and the estimation of the process variation. Commonly used subgroup sizes are between 4 and 15. Process variation is usually estimated using the average of ranges or standard deviations from subgroup samples. The performance of capability indices depends on not only the underlying distribution but also the choice of the subgroup size and the estimation of the process variation. In order to investigate the performance of C_p, C_{pu}, and C_{pk}, the following questions are asked:

1.   What is the effect of different sample sizes ?
2.   What is the effect of different process standard deviation estimates?
3.   What is the effect of the skewness of the underlying distribution?

There are various ways to investigate these questions. One way is to design an actual experiment involving these factors. Another is to develop the theoretical probability distribution of the index for a specific skewed distribution or plan a simulation study using Monte Carlo simulation techniques. It would be preferred to run an actual experiment. However, there are difficulties with this approach:

1.   The actual process cannot address different degrees of skewness.
2.   The actual process monitoring system may be difficult to adapt to study multiple sample sizes.
3.   It is costly and time consuming.

These limitations are common in many industrial processes. The theoretical approach could broaden the knowledge base but may only give results for certain restrictive distributions. Also, it is often difficult and time consuming. The Monte Carlo simulation technique is a natural choice to quickly provide some solutions for the problem. The advantages of using simulation include the following:

1.   It can be completed in a much shorter time.
2.   It is very cost effective.
3.   It can be designed to study several factors at once.
     However, simulation also has some weakness:
1.   It assumes that the quality characteristic follows a certain distribution, which may not be a realistic choice in the actual process. However, prior information or data collected previously are good sources for identifying the probability distribution of interest.
2.   It assumes the environment is constant except for those factors under study.

Simulation is commonly applied to address problems that are difficult to solve theoretically or in actual experimentation. The results are very useful to guide later actual experimentation and motivate theoretical development.

DATA AND SIMULATION PLAN

The leakage data are collected from 94 randomly selected fuel injectors. They are tested in a laboratory and automatically connected to a manifold, which is put under controlled liquid pressure, and held. For safety reasons, a nonflammable liquid is used (not the actual gasoline). The injectors are operated for a few seconds, as they would be in the engine. Leaking fluid is measured at the nozzle tip of each injector. The entire test is controlled and measured by a computer.

Name of Data File: Case19.txt

Variable Name   Description
   Case ID   Injector ID (from 1 to 94)
   Leakage   The amount of leakage (ml)

The first nine cases are as follows:

Case ID	Leakage
1	1.16
2	0.40
3	0.06
4	0.00
5	1.95
6	0.40
7	0.70
8	0.00
9	0.28

This testing procedure tries to simulate the actual operation. However, the nonflammable liquids are slightly different from gasoline in the viscosity, compressibility, and surface tension. These differences may have introduced some measurement bias of the leakage.

In planning the simulation study for evaluating the performance of capability indices, one needs to determine the following information:
A. What is the underlying distribution of the quality characteristic?
B. What are the subgroup sample sizes for the study?
C. What degree of skewness is to be compared?
D. With what are the standard deviation estimates for computing capability indices to be compared?
E. How many random samples will be used for each simulation?
F. How many simulation runs are needed?

Previously collected information can be very useful for determining the possible underlying distribution for the quality characteristic. Graphical tools such as histograms and probability plots can be used to explore the possible distribution. The leakage data suggest that if a fuel injector has leakage, then the leakage approximates a Weibull distribution well. A total of 85 out of 94 injectors have leakage. These 85 leakage data values are used for determining the leakage distribution. The reason for excluding the zero leakage injectors is due to the fact that distributions which describe nonnegative continuous characteristics usually require the characteristics to be positive. However, for the purpose of exploring the quantitative properties of a quality characteristic, these zeros should also be included.

If there is no information available for determining the underlying distribution, how should we determine the distribution of interest? In this case, since the purpose is to study the performance of indices for a highly skewed distribution, the generalized gamma distribution [Hahn and Meeker, 1991] is also a good choice.

The Weibull distribution is one of the most widely used probability distributions in the study of fracture, fatigue, strength, and reliability, etc. in industry. There are various forms of the Weibull random variable (e.g., [Hallinan Jr., 1993]). In this simulation, the two-parameter distribution is used. The scale parameter corresponds to the average leakage and the shape parameter indicates skewness of the distribution. The Weibull random variable, X, has the probability density function

$$f(x) = b(x/a)^{b-1} \exp(-x/a)^b , \, x > 0, \, a > 0, \, b > 0 ,$$

where a is the scale parameter and b is the shape parameter. When $b = 1$, the Weibull distribution is the exponential distribution. When b increases to about 3.5, the Weibull distribution is close to a normal distribution.

To determine the levels of skewness for the simulation study, the maximum likelihood method is applied to estimate the scale and shape parameters of the distribution using the 85 leakage data. The typical subgroup sample size ranges from 4 to 15. For the purpose of the comparison, the sample sizes are chosen to be 5, 8, 12, and 16.

The number of repeated samples for each simulation is analogous to the number of subgroups collected for capability analysis in the actual process. The size of simulation runs serves the purpose of investigating the distribution property of the capability index. They can be varied from several hundred to several thousand depending on the computer environment and the distribution of the statistic of interest. For this study, the number of repeated subgroup samples is chosen to be 400 and the number of simulation runs is 500.

Table 1 is the summary of five Weibull distributions for this study. The scale parameter is fixed to be one, and the shape parameter changes from 2.00 to 0.75. The corresponding mean, s.d., median, and skewness are given in Table 1. These values are obtained using the central moments of the Weibull distribution:

$$\mu_r = E(X^r) - (E(X))^r = \Gamma(r/b + 1) - \Gamma^r(1/b + 1) ,$$

where $\Gamma(.)$ is the gamma function.

Table 2 summarizes the simulation plan based on sample sizes, skewness, process standard deviation estimates, number of repeated samples, and number of simulation runs.

In planning this simulation, one needs also to determine the lower and upper specification limits (LSL and USL) for evaluating C_p, C_{pk}, and C_{pu}. For the situation when the underlying distribution is normal, the index $C_p = 1$ says that 99.73% of observations will fall within the specification limits. Equivalently, it means that LSL is the .135th percentile and USL is the 99.865th percentile. Based on the same analogy, the theoretical C_p is chosen to be one and is used as reference for the comparison. The expected LSL and USL are obtained for each underlying Weibull distribution and are summarized in Table 1.

Table 1. *Summary of the Weibull distributions for the simulation study (mean = μ_1, variance = μ_2, skewness = $\mu_3(\mu_2)^{-3/2}$).*

Parameter (Scale, Shape)	Mean	Std. Dev.	Median	Skewness	LSL (0.135 %-tile)	USL (99.865 %-tile)
(1.00 , 2.00)	0.886	0.463	0.833	0.631	0.036	2.572
(1.00 , 1.50)	0.903	0.613	0.782	1.072	0.012	3.524
(1.00 , 1.25)	0.931	0.750	0.745	1.430	0.004	4.547
(1.00 , 1.00)	1.00	1.00	0.693	2.000	0.001	6.757
(1.00 , 0.75)	1.191	1.611	0.619	3.121	0.000	14.457

Table 2. *The selected arguments for the simulation.*

Sample size (n)	5 8 12 16
Skewness	0.631, 1.072, 1.430, 2.000, 3.121
Process Std. Dev.	Use S_R , Use S_s , Use S_w , Use S_T
Number of Samples (k)	400 for each simulation run
Number of Simulations	500

The estimates of process standard deviation are four commonly used estimates. For each combination of sample size and skewness, four hundred subgroup samples are generated. Four process standard deviations are then calculated:

$S_R = \overline{R}/d_2$,

$S_S = \overline{S}/C_4$,

S_W = square root of $(s_1{}^2 + s_2{}^2 + \cdots + s_k{}^2)/k$, and

S_T is the standard deviation from the total of $n*k$ generated observations,

where $s_i{}^2$ is the ith subgroup variance. The values of d_2 and C_4 can be found in most statistical quality control books. These estimates of the process standard deviation are common in statistical quality control (e.g., [Montgomery, 1996]).

Each capability index is calculated for each combination of sample size, skewness, and process standard deviation. Five hundred simulated indices are generated for each capability index.

ANALYSIS AND SIMULATION ALGORITHM

Exploratory graphical methods are common techniques for investigating the quantitative properties and distribution of a quality characteristic. Descriptive statistics, histograms, probability plots, and box plots are used to explore the distribution of the leakage data. The Minitab statistical package is used for this case study.

Simulation study is quite different from analyzing actual data. In simulation, the most crucial steps are the planning and algorithm development. The analysis of the generated data usually is straightforward with common tools being graphical methods, descriptive analysis, and interval estimation.

For this case study, the following algorithm is used:

Step 1: Input sample size n and parameters (a,b) for the Weibull distribution.

Step 2: Generate a random sample of size n of Weibull random variates.

Step 3: Compute average, range, and standard deviation from the sample.

Step 4: Repeat Step 2 to 3 for 400 times.

Step 5: Compute four process standard deviation estimates.

Step 6: Compute C_p, C_{pu}, and C_{pk} using each of four process standard deviation estimates.

Step 7: Repeat Step 2 to 6 for 500 times.

Step 8: Compute mean and standard deviation from the 500 C_p, C_{pu}, and C_{pk}.

Step 9: Repeat Step 1 to Step 8 until all combinations of sample size and skewness are completed.

It is clear that there are many ways to develop an algorithm to carry out this study. The development of an algorithm also relies on the capability of the computer. It may take several simulations to carry out a simulation plan. Several algorithms may be needed to complete a study.

The results of the computer simulation consist of five hundred generated C_p, C_{pu}, and C_{pk} indices for a total of 80 combinations (4 sample sizes × 5 skewness × 4 process standard deviation estimates). The analysis of the simulated indices involves descriptive analysis, graphical presentation, and interval estimations.

The comparison of C_p, C_{pu}, and C_{pk} under different sample sizes can be demonstrated by multiple box plots of C_p, C_{pu}, and C_{pk} for different sample sizes. These box plots will also show the variations in C_p, C_{pu}, and C_{pk} for different sample sizes.

The comparison of C_p, C_{pu}, and C_{pk} under different skewness and process standard deviations can be demonstrated the same way.

INSTRUCTIONS FOR PRESENTATION OF RESULTS

The presentation of exploring the quantitative properties of the fuel injector leakage should include at least the following:
1. A brief description of the fuel injector process.
2. How the leakage occurs.
3. How the leakage data are collected.
4. Various graphical presentations to demonstrate the distribution of the leakage data.
5. Discuss skewed distribution and the relative standing of the average and median leakage.
6. Discussion of data transformation may also be included if the class level is appropriate.

The presentation of simulation results should include at least the following:
1. The problem statement.
2. The simulation plan and how the plan will answer the problem.
3. A summary table of the mean and standard deviation of each index generated by the simulation.
4. Various graphical presentations that will demonstrate the pattern of each index for different combinations of the factor levels.

In addition, the presentation should be understandable to nonstatisticians, especially those who are directly involved, such as operators who collect the data and managers who make decisions using the process capability analysis. How the simulation is developed is not necessary for the presentation.

REFERENCES

Devor, R., Chang, T.-H., and Sutherland, J. (1992), *Statistical Quality Design and Control*, MacMillan.

Gunter, B. (1989), The use and abuse of C_{pk}: Parts 1–4, *Quality Progress*, pp. 72–73 (January, 1989), pp. 108–109 (March, 1989), pp. 79–80 (May, 1989), pp. 86–87 (July, 1989).

Hahn, G. and Meeker, W. (1991), *Statistical Intervals—A Guide to Practitioners*, John Wiley & Sons, Inc.

Hallinan Jr., A. (1993), A review of the Weibull Distribution, *Journal of Quality Technology*, pp. 85–93.

Montgomery, D. (1996), *Introduction to Statistical Quality Control*, 3rd edition, John Wiley & Sons, Inc.

ACKNOWLEDGMENT

The authors would like to express their sincere appreciation to Roxy Peck for organizing this Academic/Industrial collaboration project and to the editor and referees for many excellent comments and suggestions. The authors are grateful for the support of both the FORD Automotive Company and Central Michigan University.

BIOGRAPHIES

Carl Lee has a BS in Agronomy, an MA in Mathematics, and received his Ph.D. in Statistics in 1984 from the Department of Statistics, Iowa State University. He has been teaching at Central Michigan University since then. He has been doing statistical consulting for academia and industry for more than ten years. Currently, he is a professor of Statistics in the Department of Mathematics, Central Michigan University, and is in charge of the statistical consulting service for the university community. His current research interest is in experimental design, industrial statistics, and statistical education. He is a member of ASA, BIOMETRICS, and ICSA.

Gus A.D. Matzo has an MS in Statistics and a BS in Chemical Engineering. He is a Staff Statistician, a Senior Technical Specialist, and an Internal Consultant primarily to Powertrain Engineering. He developed the currently used Core Competency in Statistics training materials. He has also taught numerous courses in the Industrial Engineering Department, Wayne State University. Recent publications include ORSA, SAE, and internal FORD publications. He is a member of ASA and ASQC.

CHAPTER 20

DATA FUSION AND MAINTENANCE POLICIES FOR CONTINUOUS PRODUCTION PROCESSES

Nozer D. Singpurwalla and Joseph N. Skwish

Continuous production processes involve round-the-clock operation of several, almost identical, pieces of equipment that are required to operate concurrently. Failure of one of these pieces of equipment interrupts the flow of production and incurs losses due to waste of raw material. The incidences of this in-service failure can be reduced through preventive maintenance; however, preventive maintenance also interrupts production and creates waste. Thus, the desire to prevent in-service failures while minimizing the frequency of preventive maintenance gives rise to the problem of determining an optimal system-wide maintenance interval. The aim of this study is to propose a procedure for addressing problems of this type.

INTRODUCTION

The maintenance of equipment used in continuous manufacturing processes, such as refining oil and the production of paper, steel, synthetics, and textiles, presents a generic class of problems on which little has been written. Such processes are characterized by round-the-clock operation of several pieces of almost identical equipment, called "processing stations," to which there is a continuous flow of raw material; see Fig. 1. Examples of such equipment are the spinning wheels of textile mills and the extrusion dies of chemical and steel plants.

Each processing station converts raw material to a finished product, and all stations operate concurrently. Since a common flow of raw material feeds all stations, the flow cannot be stopped to stations that are out of service. Thus, whenever a station experiences an in-service failure, there is a loss of production and a wastage of raw material.

The incidence of these in-service failures can be reduced through periodic preventive maintenance; however, since preventive maintenance also interrupts the flow of production and creates waste, it should be performed only as often as necessary. In general, the costs of stoppage due to in-service failures are much greater than those due to preventive maintenance. This is because unexpected stoppages can create an environment of chaos involving a rescheduling of maintenance personnel and a disruption of spare parts inventory.

Thus, the desire to prevent in-service failures while minimizing the frequency of preventive maintenance gives rise to the problem of determining an optimum system-wide preventive maintenance interval. The purpose of this paper is to offer a plausible solution to this problem.

The procedure we offer requires as input two quantities. The first is a probability model for the failure of the equipment; the second is a utility function which describes the consequences of scheduled and unscheduled stoppages. The proposed failure model is based on expert opinion and the pooling, or the fusion, of data from the various pieces of equipment. The fusion of data is based on the information content of each data set in the sense of Shannon.

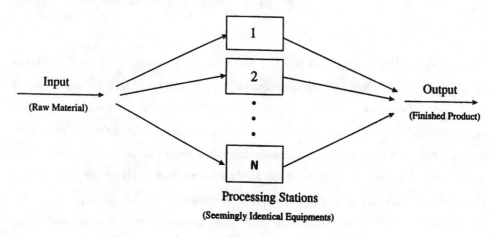

Fig. 1. *A continuous production process with a common flow of raw material.*

BACKGROUND INFORMATION

Our approach to the decision problem described above is based on the principle of maximization of expected utility (see [Lindley, 1985]). An implementation of this principle requires the specification of two quantities: (i) a utility function, which relates the profits of production to the losses of stoppage, and (ii) a probability model for the equipment's failure time. In this section we describe a piecewise linear utility function appropriate for use with continuous production processes. This function, or its nonlinear versions, are flexible enough for multipurpose use. We also discuss the development of an "omnibus" failure model, that is, a model encompassing all of the almost identical pieces of equipment.

The omnibus, or "composite" failure model (see [Chen and Singpurwalla, 1996]) is developed in two stages. In the first stage an individual model for each piece of equipment in the system is developed. Each model is developed via an approach which incorporates expert opinion and the failure/survival history for that piece of equipment. In the second stage, the individual models are fused to obtain an omnibus model. The fusion is performed through a pooling formula, in which the weights assigned to the individual models are based on the Shannon information (see [Lindley, 1956]) provided by the failure/survival history of each piece of equipment. Our reason for a two-stage construction of the omnibus model is that, although the equipments perform identical functions and are often of the same design, they differ from each other with respect to age, maintenance history, and operating environment. Thus, from a statistical point of view, they cannot be considered identical.

When solving decision problems, the choice of a utility function is often the subject of debate and discussion. Thus, the need to have utility functions that are representative of a certain class of applications is germane. The class of applications of interest to us are continuous production processes having several processing stations, all serviced by a common flow of raw material. Such features mandate that the utility function be continuous and increasing with time whenever production is in progress, and be decreasing whenever there is a stoppage. The shape of the utility function dictates the rate at which there is a gain or a loss of utility.

In the utility function we have chosen, the profit (positive utility) for each processing station is a linearly increasing function of production time t, with a slope determined by an angle, say $\theta(t^*)$, which is a function of t^*, the time at which the production stops. Thus, if t is an optimal maintenance interval, and if the processing station experiences no in-service failure, then the profit until t is $\tan\theta(t) \cdot t$; see Fig. 2. At time t, preventive maintenance is performed, and if x is the time needed to complete a scheduled maintenance action, then the loss (negative utility) due to stoppage, to include the cost of wasted material, is $\tan\theta_1 \cdot x$, where $\theta_1 > \theta(t)$; both θ_1 and $\theta(t)$ have to be specified. Added to this loss is an amount c, the cost of actually performing a scheduled maintenance; see Fig. 2. In Fig. 3 the case of a processing station experiencing an in-service failure at t^* is illustrated. The profit due to production is $\tan\theta(t^*) \cdot t^*$, but the loss of an unscheduled stoppage is $\tan\theta_1 \cdot kx$, where the constant k (>1) reflects the incremental consequences of an unscheduled maintenance. Again added to this cost is the amount c. There are possible generalizations to the set-up of Fig. 3, the obvious ones being that both k and c can be made decreasing functions of t^* with $k(t) = 1$, and $c(t) = c$. Also, the utility function need not be linear. For $\theta(t^*)$ a plausible choice is $\theta(t) \cdot (t^* / t)$; this choice reflects the feature that small failure times result in a decrease in the rate of profit from production.

If T denotes the time to failure of any processing station in the system, and if an omnibus failure distribution for T has density $f(t)$ and survival function $S(t \mid H) = P(T \geq t \mid H)$, then $U(t)$, the expected utility per processing station, for a production run that is designed to be of length t, is of the form

$$(1) \qquad U(t) = S(t \mid H)\left[\tan\theta(t) \cdot t - \tan\theta_1 \cdot x - c\right] + \int_0^t \left\{\tan\theta(t^*) \cdot t^* - \tan\theta_1 \cdot kx - c\right\} f\left(t^* \mid H\right) dt^*,$$

where the H associated with f and S denotes history, which includes such things as background information and model parameters, and $\theta(t^*) \cdot (t^* / t)$, with $\theta(t)$ specified. Once f and S are specified, $U(t)$ can be evaluated and optimized.

We now describe a two-stage procedure for specifying a common distribution to characterize the failure behavior of any processing station in the system. The first stage of the procedure involves the development of a failure model for each station. The approach used is Bayesian; it involves the incorporation of expert opinion and the failure/survival history of each station. The second stage consists of pooling the individual models to obtain a composite model.

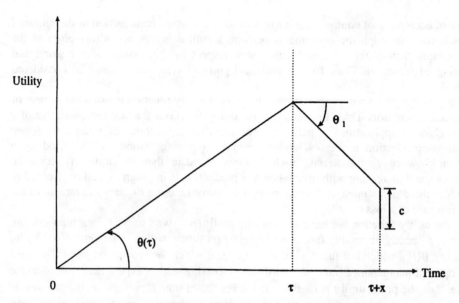

Fig. 2. *Utility function without an in-process failure.*

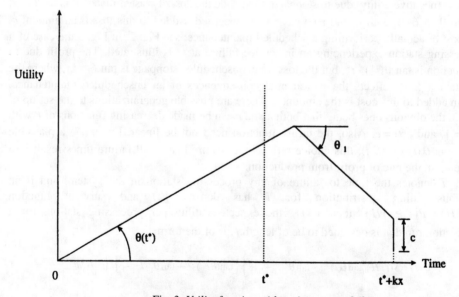

Fig. 3. *Utility function with an in-process failure.*

A Bayesian approach for developing failure models based on expert opinion and life data has been proposed by Singpurwalla (1988). This approach has been codified for use on a personal computer; see [Aboura and Campodónico, 1992]. It is based on the assumption that the time to failure distribution is a member of the Weibull family. That is, if T_i denotes the time to failure of the ith processing station, $i = 1, 2,...,N$, then, given a scale (shape) parameter $\alpha(\beta)$, the probability density of T_i at t is

(2) $$f(t|\alpha, \beta) = (\beta / \alpha^\beta)t^{\beta-1}e^{-(t/\alpha)\beta}, \quad t > 0.$$

A prior distribution for α and β is assessed via consultation with an expert who is able to provide an opinion about the shape parameter β and the median M_i of T_i, where $M_i = \alpha \exp(c/\beta)$, with $c = \ln(\ln 2)$. Let m_i denote the expert's declared value for the most likely value of M_i, and let s_i be the expert's measure of uncertainty about m_i. Since the parameter β characterizes the aging behavior of processing station i, opinion about β is elicited from the expert in terms of λ_i and p_i, the scale and the shape parameters, respectively, of a gamma distribution for β. Recall that the mean and the variance of a gamma distribution with scale parameter λ and shape parameter p are p/λ and (p/λ^2), respectively. Once the hyperparameter $\Theta_i = (m_i, s_i, \lambda_i, p_i)$ is elicited from the expert, then, assuming independence of M_i and β, a joint prior distribution of α and β can be easily induced. The details are in [Singpurwalla, 1988], where issues such as the expertise of the expert, the case of multiple experts, and correlations between experts are also discussed. Let the density, at α and β, of the induced joint prior distribution of α and β for the ith processing station be denoted by $\pi_i(\alpha, \beta | \Theta_i)$.

Suppose that there is a maintenance schedule in place for the production system; this schedule need not be optimal. Then, a life history of each processing station will consist of both failure and survival times. Let x_{ij} and s_{ih} denote the n_i failure times and the r_i survival times, respectively, of the ith station, where $j = 1,2,...,n_i$ and $h = 1,2,...,r_i$. Let $D_i = (x_{ij}, s_{ih}; j = 1,...,n_i; h = 1,...,r_i)$ denote the lifetime history of the ith station, $i = 1,..., N$. Then the likelihood of α and β, $L(D_i; \alpha, \beta)$, can be easily written from which the posterior distribution of α and β, given D_i and Θ_i, can be obtained via an application of Bayes' Law as

(3)
$$\pi_i(\alpha, \beta | D_i, \Theta_i) \infty L(D_i; \alpha, \beta) \pi_i(\alpha, \beta | \Theta_i).$$

While a knowledge of the posterior distributions of α and β is of interest, especially that of β, what we really need to know is the posterior survival function of T_i,

$$S_i(t | D_i, \Theta_i) = P(T_i \geq t | D_i, \Theta_i).$$

For this, we invoke the law of total probability, from which

(4)
$$S_i(t | D_i, \Theta_i) = \int_{(\alpha, \beta)} P(T_i \geq t | \alpha, \beta) \pi_i(\alpha, \beta | D_i, \Theta_i) d\alpha d\beta ,$$

where $P(T_i \geq t | \alpha, \beta) = \exp(-(t/\alpha)^\beta)$. Analogously, the prior survival function of T_i is

(5)
$$S_i(t | \Theta_i) = \int_{(\alpha, \beta)} P(T_i \geq t | \alpha, \beta) \pi_i(\alpha, \beta | \Theta_i) d\alpha d\beta .$$

Because the expression for $\pi_i(\alpha, \beta | \Theta_i)$ is not available in closed form, (4) and (5) will have to be evaluated numerically. These computations are performed by IPRA (see [Aboura and Campodónico, 1992]) once the inputs Θ_i and D_i, $i = 1,..., N$, are provided.

At this point we have at our disposal N survival functions, $S_i(t|D_i, \Theta_i)$, $i = 1,...,N$, one for each of the N stations. Our task is to pool (or combine) these survival functions to obtain an omnibus survival function $S(t|D, \Theta)$, where $D = (D_1,..., D_N)$, and $\Theta = (\Theta_1,..., \Theta_N)$. We now propose a procedure for accomplishing this task.

The pooling of distributions is a well-discussed topic in the statistical literature; a recent overview, with new insights, is in Dawid, DeGroot, and Mortera (1995). The majority of pooling procedures that have been proposed involve the concept of a decision maker (DM), or a "boss," who consults several experts j, $j=1,2,...$, regarding their assessments about the same unknown quantity. The DM then combines all the experts' inputs based on the DM's judgment about the expertise of each expert, and on the DM's perceived correlations between their inputs. Indeed, our procedure for inducing a prior distribution for α and β of processing station i is based on the notion of a DM eliciting information from a single expert specializing on station i. The DM's judgments about the expertise of an expert, if any, will be reflected in the prior through an expanded version of the hyperparameter Θ_i.

It is, of course, conceivable that there could be several experts specializing on a processing station and that there could be experts specializing on more than one station. Thus, using many of the available procedures for pooling survival functions could entail a duplicate use of the DM's judgments about the expertise of the experts. Our strategy for pooling avoids such duplications; moreover, it accounts for the fact that the life history data of each processing station is of different dimensions. For instance, some processing stations may have been more carefully scrutinized than others, and some may have experienced more failures than others. How do we account for the fact that a processing station with 1 failure and 99 survivals could be more or less informative (about the reliability) than another station which has experienced 0 failures and 9 survivals? That is, how should our pooling procedure take into account the fact that data set D_i could be more informative than data set D_j? The situation here is reminiscent of an observation by Abel and Singpurwalla (1994) that survivals could be more informative than failures, and vice versa.

The procedure for pooling that we propose is the "linear opinion pool" (discussed, for example, by [Genest and Zidek, 1986]). The linear opinion pool is simply a weighted linear combination of the N survival functions, $S_i(t|D_i, \Theta_i)$. The difficulty in implementing a linear pool is the choice of the weights w_i, $i = 1,..., N$, that are to be assigned to each survival function. Our strategy is to weigh each survival function in proportion to the "observed Shannon information" (cf. [Singpurwalla, 1996]) about T_i that is provided by D_i, the data. This pooling strategy is ad hoc; its formal justification remains to be provided. The issue of justification is important because there are philosophical difficulties with the linear opinion pool (cf. [Genest and Zidek, 1986]).

Given Θ_i, the *observed information* about T_i provided by the data D_i is defined as

$$(6) \qquad I(T_i|D_i, \Theta_i) = \int_S \phi(s|D_i, \Theta_i) \ln(\phi(s|D_i, \Theta_i)) ds - \int_S \phi(s|\Theta_i) \ln(\phi(s|\Theta_i)) ds ,$$

where $\phi(s|D_i, \Theta_i)$ is the density at s of the posterior distribution of Ti; that is,

$$\phi(s|D_i, \Theta_i) = -d/dt[S_i(t|D_i, \Theta_i)].$$

Similarly, $\phi(s|\Theta_i)$ is the density of the prior of T_i. The weight w_i to be assigned to $S_i(t|D_i,\Theta_i)$ is

(7) $$w_i = I(T_i|D_i,\Theta_i)/\sum_{j=1}^{N} I(T_j|D_j,\Theta_j), \quad i=1,...,N.$$

An omnibus survival function that we have been seeking is then given as

(8) $$S(t|D,\Theta) = \sum_{i=1}^{N} w_i S_i(t|D_i,\Theta_i).$$

Note that the w_i's are the normalized Kullback–Liebler distance between the prior and the posterior.

QUESTION OF INTEREST

Specifically, the question that we want to answer is: What is the optimal system-wide maintenance interval for a continuous production process?

DATA

Data files: Case20A.txt, Case20B.txt, Case20C.txt

Variable Name	Description
Number of Failures	n_i
Failure Times	$x_{i\,i}$
Number of Survivals	h_i
Survival Times	$s_{i\,i}$
Observed Information	See equation (6)
Pooling Weights	See equation (7)

Table 1. *Failure times of the five processing stations.*

	Station 1	Station 2	Station 3	Station 4	Station 5
Number of Failures	13	13	11	8	4
Failure Times	7.72	5.08	13.94	13.92	23.19
	25.20	16.87	20.87	6.21	13.28
	•••				

Table 2. *Survival times of the five processing stations.*

	Station 1	Station 2	Station 3	Station 4	Station 5
Number of Survivals	10	11	12	14	14
Survival Times	19.92	18.98	17.33	14.73	13.26
	13.43	15.88	0.06	13.98	15.25
	•••				

Table 3. *Observed information and the pooling weights.*

	Station 1	Station 2	Station 3	Station 4	Station 5
Observed Information	0.864	0.454	0.389	0.817	0.966
Pooling Weights	0.247	0.130	0.112	0.234	0.277
	•••				

ANALYSIS

To illustrate an application of the material of the previous sections to a real-life scenario, we consider a continuous production process with 5 concurrently operating processing stations in a chemical plant. Ideally, the system is required to function 24 hours per day, 7 days per week. Current operating procedures call for a preventive maintenance action after a specified number of hours of operation. There is, however, no assurance that these procedures are strictly followed. The 5 stations were observed for a period of 2 weeks providing the failure and survival times shown in Tables 1 and 2, respectively; for reasons of company confidentiality, these times are transformed via a common location and scale parameter.

An engineer, considered to be an expert on the system, was consulted for his assessment about the median lifetimes and aging characteristics of each processing station. His judgment was that

- The processing stations neither degrade nor improve with use;
- The most likely value of a station's median life is 24 (camouflaged) hours;
- The expert's assessed standard deviation for the assessed median is 6 hours (also transformed).

Since we have no basis to assume that the expert's assessments were biased, there was no need to modulate the expert's inputs, so we set $m_i = 24$, $s_i = 6$, $\lambda = 2$, and $p = 1$ for $i = 1,...,5$. These values were used in IPRA which, with the data of Tables 1 and 2, produced the 5 survival curves shown in Fig. 4. To pool these 5 survival functions, we computed the observed information provided by each set of data, see Table 3, and the ensuing pooling weights. These weights were then used to obtain the omnibus (posterior) survival function shown in boldface in Fig. 4.

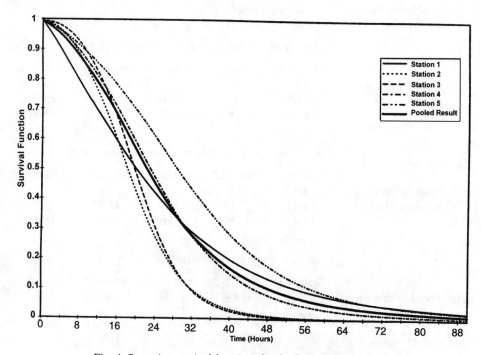

Fig. 4. *Posterior survival functions for the five processing stations.*

Observe that even though stations 2 and 3 contain approximately the same number of failures and survivals as stations 1 and 4, the weights that they receive are less than those given to the others. Furthermore, station 5, which has the smallest number of failure, receives the largest weight. What could be the possible reasons for this asymmetry? One explanation is that the information provided by the data about T_2 and T_3, the failure times of stations 2 and 3, may not be too different from that subsumed in the prior distribution (vis-à-vis the expert inputs), as compared to the information provided by the data about T_1, T_4, and T_5. Another explanation pertains to the scatter of the observations (both failures and survivals) for the 5 stations; it has different patterns. Thus, what matters is not only the numbers of survivals and failures, but what failure and survival times are observed. Note that the weights would be very difficult to assess by subjective considerations alone.

CONCLUSION

Replacing the $f(t|H)$ of equation (1) by $S(t|D,\Theta)$ of equation (8), we have the expected utility, per processing station, per production cycle, given a maintenance interval t as

$$U(t)= S(t|D,\Theta)[tan\,\theta(t)\cdot t - tan\,\theta_1\cdot x - c]+\left[\int_0^t \{tan\,\theta(t^*)\cdot t^* - tan\,\theta_1\cdot kx - c\}\cdot f(t^*|D,\Theta)dt^*\right],$$

where

$$f(t|d,\Theta)=-\frac{d}{dt}S(t|D,\Theta).$$

Recall that we have chosen $\theta(t^*)=\theta(t)\cdot(t^*/t)$, where t is the preventive maintenance interval, and $\theta(t)$ is specified. Also specified are the quantities θ_1, x, c, and k, with $\theta_1 > \theta(t)$ and $k>1$. The optimal maintenance interval t is that value of t which maximizes $U(t)$.

Since $S(t|D,\Theta)$ is not available in closed form, the maximization of $U(t)$ with respect to t has to be done numerically. This requires that we choose a grid of values of t, say t_1, $t_2,...,t_u,...$, and for each value in the grid evaluate $U(t)$. Thus, for $t=t_u$, we evaluate $U(t_u)$ by using $S(t_u|D,\Theta)$, and approximating $f(t_u|D,\Theta)$ by

$$-[S(t_{u+1}|D,\Theta)-S(t_u|D,\Theta)]/(t_{u+1}-t_u),$$

the negative of the slope of $S(t|D,\Theta)$. The finer the grid, the better the resulting approximation. We then choose τ as that value of t in the grid for which $U(t)$ is a maximum. An alternative approach is to approximate $S(t|D,\Theta)$ by a well-known function and proceed analytically. Thus, for example, if $S(t|D,\Theta)$ is approximated by the function $e^{-\lambda t}$, with λ chosen so that $S(t|D,\Theta)$ and $e^{-\lambda t}$ are as close to each other as is possible, then the maximization of $U(t)$ is relatively straightforward, especially when $\theta_1 = \theta(t)$. Thus, for example, if $x = 1/3$, $c = 1$, $k = 3$, and $\lambda = 1/12$, then $U(t)$ attains its maximum at $t \approx 20$ for a range of values of θ; see Fig. 5, which shows the behavior of $U(t)$ for $\theta(t) = 10, 20, 30$, and 40. When $\lambda = 1/24$, $U(t)$ attains its maximum in the vicinity of $t=42$, as the plots of Fig. 6 show. Thus, it appears that the maximum of $U(t)$ is quite sensitive to the choice of λ (and hence to $S(t|D, \Theta)$) and relatively insensitive to $\theta(t)$. Figures 7 and 8 are a magnification of Fig. 6, in the vicinity of $t = 42$, for $\theta(t) = 10$ and $\theta(t) = 40$, respectively. Figures 5 and 6 illustrate the crucial role of the omnibus survival function in arriving at an optimal maintenance interval. Clearly, the development of the omnibus survival function needs to be treated with a careful deliberation.

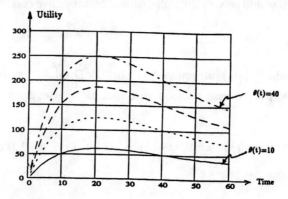

Fig. 5. Behavior of the utility function for different values of $\theta(t)$ when $\lambda = 1/12$.

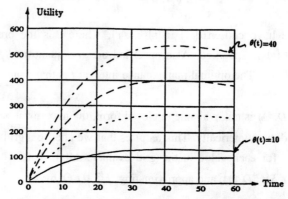

Fig. 6. Behavior of the utility function for different values of $\theta(t)$ when $\lambda = 1/24$.

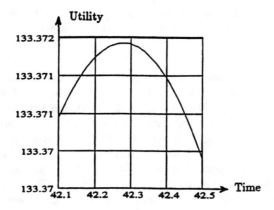

Fig. 7. *A magnification of Fig.* 6 *when* $\theta(t)=10$, *near* $t=42$.

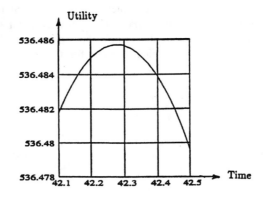

Fig. 8. *A magnification of Fig.* 6 *when* $\theta(t)=40$, *near* $t=42$.

REFERENCES

Abel, P. S. and N. D. Singpurwalla (1994), *To Survive or to Fail: That Is the Question,* The American Statistician, 48 1:18–21.

Aboura, K. and S. Campodónico (1992), *IPRA, An Interactive Procedure for Reliability Assessment Incorporating the Opinion of One or Two Experts and Failure/Survival Data, User's Manual,* Technical report GWU/IRRA/Serial TR-92/3, Institute for Reliability and Risk Analysis, The George Washington University, Washington, DC.

Chen, J. and N. D. Singpurwalla (1996), *Composite Reliability and Its Hierarchical Bayes Estimation,* Journal of the American Statistical Association, 91, 436:1474–1484..

Dawid, A. P., M. DeGroot, and J. Mortera (1995), *Coherent Combination of Experts' Opinions,* Test, 4 2:263–313.

Genest, C. and J. V. Zidek (1986), *Combining Probability Distributions: A Critique and an Annotated Bibliography,* Statistical Science, 1:114–148.

Lindley, D. V. (1956), *On a Measure of Information Provided by an Experiment,* Annals of Mathematical Statistics, 27:986–1005.

Lindley, D. V. (1985), *Making Decisions,* Second Edition, New York: John Wiley & Sons.

Singpurwalla, N. D. (1988), *An Interactive PC-Based Procedure for Reliability Assessment Incorporating Expert Opinion and Survival Data,* Journal of the American Statistical Association, 83 401:43–51.

Singpurwalla, N. D. (1996), *Entropy and Information in Reliability,* Bayesian Analysis in Statistics and Econometrics, New York: John Wiley & Sons.

BIOGRAPHIES

Nozer D. Singpurwalla is Professor of Operations Research and of Statistics and Distinguished Research Professor at The George Washington University in Washington, DC. He has been Visiting Professor at Carnegie-Mellon University, Stanford University, the University of Florida at Tallahassee, and the University of California at Berkeley. He is a Fellow in The Institute of Mathematical Statistics, the American Statistical Association, and the American Association for the Advancement of Science, and he is an elected member of the International Statistical Institute. He has coauthored a standard book in reliability and has published over 135 papers on reliability theory, Bayesian statistical inference, dynamic models and time series analysis, quality control, and statistical aspects of software engineering.

After earning his B.S. in Engineering Science from Pennsylvania State University, Joseph N. Skwish worked for Eastman Kodak while earning an M.S. in Industrial Statistics at the University of Rochester. Since completing his Ph.D. in Statistics at John Hopkins in 1969 he has been employed at the DuPont Company as a statistical consultant. He is a fellow of the American Statistical Association and a member of the Institute of Mathematical Statistics and the American Society for Quality.

Index